Veterinary
Office Practices

Join us on the web at

Agriscience.delmar.cengage.com

Veterinary Office Practices

EDITED BY

Robert Kehn

DELMAR
CENGAGE Learning·

Australia · Brazil · Japan · Korea · Mexico · Singapore · Spain · United Kingdom · United States

DELMAR
CENGAGE Learning™

Veterinary Office Practices
Robert Kehn

Vice President, Career Education Business
    Unit: Dawn Gerrain

Director of Editorial: Sherry Gomoll

Acquisitions Editor: Zina M. Lawrence

Developmental Editor: Andrea Edwards

Editorial Assistant: Rebecca Switts

Director of Production: Wendy A. Troeger

Production Manager: Carolyn Miller

Production Editor: Kathryn B. Kucharek

Director of Marketing: Donna J. Lewis

Channel Manager: Nigar Hale

Cover Image: PhotoDisc

Cover Design: Panarama Design

For product information and technology assistance, contact us at
Cengage Learning Customer & Sales Support, 1-800-354-9706

For permission to use material from this text or product,
submit all requests online at **www.cengage.com/permissions**
Further permissions questions can be emailed to
permissionrequest@cengage.com

Library of Congress Control Number: 2003041464

ISBN-13: 978-1-4018-1569-1

ISBN-10: 1-4018-1569-3

Delmar
Executive Woods
5 Maxwell Drive
Clifton Park, NY 12065
USA

Cengage Learning is a leading provider of customized learning solutions with office locations around the globe, including Singapore, the United Kingdom, Australia, Mexico, Brazil, and Japan. Locate your local office at
international.cengage.com/region

Cengage Learning products are represented in Canada by Nelson Education, Ltd.

For your lifelong learning solutions, visit **delmar.cengage.com**

Visit our corporate website at **www.cengage.com**

Printed in Canada
10 11 12 13 11 10 09

Contents

Preface

Veterinary Office Practice is a required course in veterinary assisting, veterinary technician, and pre-DVM (Doctor of Veterinary Medicine) programs taught at career, community, and 4-year colleges. This course covers the business and professional aspects of a practice, including ethical and legal considerations, client communications, public relations, accounting, scheduling, and record keeping. This book, *Veterinary Office Practices*, has been developed to meet this need, providing an overview of the duties and responsibilities of veterinary team members in the practice, with a focus on the dual medical/administrative responsibilities of veterinary assistants.

ORGANIZATION OF THE TEXT

Veterinary Office Practices is divided into three parts that address the practice environment, the veterinary team member, and financial issues that complement the standard format of the Veterinary Office Practice course.

Part I addresses the veterinary office itself, including descriptions of veterinary office settings and the various members of a typical veterinary staff, the veterinary practice facility itself, including facility care and maintenance, administrative duties such as keeping and filing medical records and making appointments, and the role of computers in veterinary practice. Part II focuses on the individual and what a person needs to do to communicate effectively; how to manage, prevent, and cope with stress; how to interact with clients in person and on the phone; concerns related to euthanasia and grief management; and ethics in veterinary practice. Finally, Part III addresses financial issues, including a general overview of veterinary fees, collection procedures, billing, and payroll accounting.

FEATURES

Veterinary Office Practices has been organized so that it not only provides factual information, it also facilitates study and review with features such as:

- *Chapter Objectives.* Students can use the list of objectives that accompanies each chapter to help them understand what they should learn from the reading and where to focus their attention.
- *Key Terms.* The most important terms in each chapter are identified in a list at the outset, and then identified in bold type when they first appear in the text.
- *Chapter Summaries.* A brief synopsis is included at the end of each chapter, re-presenting the main points and helping students to tie the material together.
- *Review Questions.* Each chapter includes a list of review questions that test students' comprehension of the material, helping them to determine how well they understand what they read.
- *Case Studies.* Students learn better when they can apply concepts explained in a text to a real-world setting. With this in mind, case studies exhibiting the practical application of instructions and information supplied in each chapter have been included to aid students' comprehension.
- *On-line Resources.* Each chapter also includes recommended on-line resources, including links to both specific topics addressed in the text and related content available on the Internet.

INSTRUCTOR'S GUIDE

Also offered with this book is the *Instructor's Guide to Accompany Veterinary Office Practices*, a chapter-by-chapter companion with classroom aids, featuring:

- answers to case study questions
- answers to review questions
- classroom exercises
- test bank questions and answers

ABOUT THE EDITOR

Robert Kehn is an experienced freelance researcher, writer, and editor with experience developing projects in the education, transportation, public health, technology, arts, management consulting, and legal industries. His previous work with Delmar includes WebTutor™ projects in early childhood education, law, esthetics, and transportation. He also edited the *Adams Internet Job Search Almanac, Sixth Edition* in 2002. Mr. Kehn lives in Troy, New York, and holds a degree in English from the State University of New York at Albany.

ACKNOWLEDGMENTS

The author and Delmar wish to thank the following reviewers for their time and content expertise:

Susan Burnett, D.V.M.
Argosy University
Bloomington, MN

Darwin Yoder, D.V.M.
Sul Ross State University
Alpine, TX

Sheryl Keeley, C.V.T.
Northwestern Connecticut Community College
Winsted, CT

Betsy Krieger, D.V.M.
Front Range Community College
Fort Collins, CO

Dr. Stuart L. Porter
Blue Ridge Community College
Weyers Cove, VA

INTERNET DISCLAIMER

The authors and Delmar affirm that the Web site URLs referenced herein were accurate at the time of printing. However, due to the fluid nature of the Internet, we cannot guarantee their accuracy for the life of the edition.

PART I

The Veterinary Office

UNIVERSITY OF NEBRASKA - LINCOLN
Institute of Agriculture & Natural Resources
Department of Agricultural Leadership
Education and Communication
P.O. Box 830709
Lincoln, NE 68583-0709

CHAPTER 1

Introduction to Veterinary Practice

OBJECTIVES

When you complete this chapter, you should be able to:

- identify and describe the different staff roles within a typical veterinary practice and the professional organizations associated with them

- identify and describe different types of veterinary care facilities

- describe useful administrative techniques

- explain what cross-training is and how it can benefit both the individual and the practice

- explain the purpose and importance of an office procedures manual and a personnel manual

KEY TERMS

veterinarian
American Veterinary Medical Association (AVMA)
veterinary technician
National Association of Veterinary Technicians in America (NAVTA)
veterinary assistant

American Veterinary Assistants Association (AVAA)
receptionist
bookkeeper
practice manager

American Animal Hospital Association (AAHA)
corporate veterinary medicine
cross-training
personnel manual
procedures manual

INTRODUCTION

Practice management has become a critical factor in the success or failure of veterinary practices. Sound management techniques have their place in every veterinary practice, whether it is a fully staffed veterinary hospital, a companion animal practice with five veterinarians, or a livestock animal practice with one doctor making farm calls.

When you think of a veterinary practice, you usually think first of the veterinary health care team providing medical care to dogs, cats, horses, or cattle. Veterinary care may be delivered in an established clinic or hospital, in a client's living room on a house call, or in a barnyard on a farm call. However, delivering medical care is only one aspect of what is involved in a veterinary practice (Figure 1-1). Much must happen behind the scenes in order for the veterinarians and staff to be able to provide the care demanded of them.

Coordination of these behind-the-scenes issues is at the heart of veterinary practice management. All members of the veterinary health care team are, by definition, involved in practice management issues, although "practice manager" may not be part of their title or job description.

STAFFING THE PRACTICE

Who are the individuals involved in a typical veterinary practice? Obviously, the first person that comes to mind is the **veterinarian**. The veterinarian is a

FIGURE 1-1   Your duties as a veterinary assistant may include more than just assisting the veterinarian with direct medical care. *(Setting courtesy of Airpark Animal Hospital, Westminster, MD)*

FIGURE 1-2 The veterinary technician may be responsible for performing diagnostic blood tests.

doctor that specializes in the treatment of animals, including companion animals, livestock, zoo animals, sporting animals, and laboratory animals. He is responsible for diagnosing illnesses, vaccinating against diseases, prescribing medications, performing surgery, treating wounds and broken bones, and advising clients about nutrition, behavior, and other health-related concerns. To become a veterinarian, a person must earn a Doctor of Veterinary Medicine (DVM) from an accredited college, and pass a state exam to obtain a license to practice. Continuing education is required for veterinarians to maintain their licenses, and many are members of the **American Veterinary Medical Association** (AVMA), a professional association dedicated to advancing the field of veterinary medicine and all of its related aspects.

Most veterinary practices employ at least one certified or licensed **veterinary technician**. The technician is a trained professional, much like a nurse in human medicine. He works closely with the veterinarian in the delivery of medical care. The technician is often responsible for educating clients, treating patients within the hospital, administering medications (including anesthesia), and performing diagnostic techniques such as laboratory analysis of blood (Figure 1-2). He may also perform physical examinations, take patient histories, and assist in surgery. Many veterinary technicians are also involved in some degree of practice management, depending on the size and organization of the office or hospital. They also have their own organization, the **National Association of Veterinary Technicians in America** (NAVTA), dedicated to fostering high standards of veterinary care, promoting the health care team, and advancing the veterinary technology profession.

To become a veterinary technician, a person must earn a minimum of an associates degree, and then become licensed through his state. Because they are licensed, technicians are required to attend continuing education seminars, classes, and other educational opportunities to maintain and renew their licenses.

There is another level of veterinary technician called *veterinary technologist*. The technologist must earn a four-year degree in an accredited veterinary technology program, and then go through the state licensing program. Continuing education is also required for veterinary technologists.

You will also often find **veterinary assistants** who assist veterinary technicians and veterinarians. As a veterinary assistant, you may be asked to restrain an animal while the veterinarian performs a physical examination or draws a blood sample, or to assist in the nursing care of hospitalized animals. Veterinary assistants may also aid the veterinarian in surgery, administer transfusions and intravenous fluid therapy, give injections (as directed by the veterinarian), apply bandages and splints, place and maintain urinary catheters, and sterilize instruments. Additionally, veterinary assistants are often responsible for patient education, filling prescriptions, performing certain laboratory procedures, bathing and exercising animals, answering phones, scheduling appointments, and helping to keep the practice clean. To become a veterinary assistant, you generally either earn certification through a school offering a veterinary assistant program, or you are trained on the job by a veterinarian. There is currently no licensing program for assistants, although they do have their own national organization, the **American Veterinary Assistants Association** (AVAA), dedicated to bringing recognition and respect to the veterinary assistant profession.

Although technicians and assistants provide a certain level of patient care and treatment, this must always be done under the supervision of the veterinarian. *Only the veterinarian can diagnose conditions, prescribe medications, and perform surgery.*

In addition to those individuals delivering medical care, a veterinary practice or hospital usually includes a **receptionist** who answers the phone, greets clients as they enter, coordinates the scheduling of patient appointments, and otherwise oversees the workings of the front desk area. Most practices and hospitals today have a **bookkeeper**, or even an accountant, who takes care of the accounting aspects of the practice, including payroll, paying bills, and ordering supplies.

Many veterinary practices are now employing an individual whose specific duties are in the area of veterinary practice management. This person's title may be *clinic manager*, **practice manager**, or *business manager*. Although all members of the veterinary health care team contribute to the smooth running of the veterinary practice, this individual's job is to oversee and coordinate the behind-the-scenes tasks and duties that allow efficient delivery of medical care to patients and the accommodation of clients. In many cases, the practice manager is also a licensed or certified veterinary technician.

VETERINARY SETTINGS

Veterinary medicine is practiced in a variety of settings, including single-veterinarian practices, multiveterinarian practices, animal hospitals, zoos, and specialty practices such as poultry or equine.

Single-veterinarian and multiveterinarian practices are quite common. Most clients taking their pets in for health care will visit one of these types of practices. In the single-veterinarian setting, the veterinarian is the leader of the veterinary health team, and is ultimately responsible for all that happens in the practice. In a multiveterinarian practice, the veterinarians may share equal responsibility or one veterinarian may act as the leader of the team. In either case, the veterinarian always has greater authority over patient care than the veterinary technician or assistant. Many veterinarians also make house calls to farms, zoos, or other places with large animals or large numbers of animals, where it would be impractical to bring the animal(s) to the practice or hospital.

An animal hospital may provide general health services that a standard practice offers, but it is also designed to handle all-hours emergency care and hospitalization for a greater number of animals. Veterinarians that work in animal hospitals are often members of the **American Animal Hospital Association** (AAHA), an organization dedicated to professional veterinary development, quality hospital standards, and excellence in the delivery of veterinary medicine.

Pharmaceutical companies, research institutes, zoos, and humane societies. often employ veterinarians or veterinary teams that specialize in treating specific types of animals. These veterinary services provided outside of the traditional practice/hospital setting are collectively referred to as practitioners of **corporate veterinary medicine**.

THE ROLE OF THE VETERINARY ASSISTANT IN PRACTICE MANAGEMENT

As a veterinary assistant, your role in the management of the veterinary office will vary dramatically from practice to practice. You may have no occasion to write an invoice or complete a transaction with a client. You may never be asked to keep track of the office supplies used in your practice. In contrast, you may be asked routinely to participate in the business office activities of the practice where you work. Perhaps you will be asked to cover the front desk while the receptionist takes a lunch break (Figure 1-3). You may be asked to file medical records that have been appropriately processed. It is up to you to maintain an open mind about doing something, when asked, that does not directly involve working with animals.

FIGURE 1-3   Your duties may include covering the front desk area while the receptionist takes a break. *(Setting courtesy of Airpark Animal Hospital, Westminster, MD)*

Time Management

One of the first tasks you will face in becoming acquainted with the business aspects of a veterinary practice is to learn how time will flow throughout the day. What is the daily routine? You must become comfortable with many important details of time management within a veterinary practice. You will not want to overlook any of them.

To help you get acquainted with the daily flow of activity in the veterinary practice where you work, seek answers to the following questions:

- During what hours is the practice open to see clients and patients?
- Does the practice close for lunch?
- During what hours is the doctor (or doctors) scheduled to see animals?
- Are patients seen on a walk-in basis, or are appointments preferred?
- Does the practice offer house or farm calls?
- How and when are house calls and farm calls scheduled? Are there specific time slots during the day for these types of appointments, or are they simply scheduled according to clients' requests?
- On what days are surgical and dental procedures performed?
- At what time of day are surgical and dental procedures performed?
- What are my hours as veterinary assistant?
- When should I take my lunch break?
- Will I work only on the premises, or will I occasionally be asked to assist on a house or farm call? (This is important if you are going to have

tasks to perform on your own during the day. You will be more able to plan your time for those tasks if you know whether your presence is required outside the practice.)

When considering time management within a veterinary practice, the central issue involves the scheduling of patient appointments. You will want to know how the appointments are scheduled and who schedules them. If only the receptionist makes appointments, you will not need to be quite as detailed in your understanding of the scheduling. If, however, you will be making appointments for clients, you will need to have the answers to the following questions:

- If a client asks for a particular veterinarian in a multidoctor practice, when is that doctor available to see patients (to avoid a conflict with performing surgery or having a day off from the practice)?
- How and when are surgeries scheduled? Are there specific surgery days? How many surgeries are scheduled per day?
- What is the client told about preparing for surgery (i.e., time to arrive, keeping food from his animal for an appropriate time, time to pick up the animal from the practice)?
- Who will discharge the patient? (In some practices, every surgery is discharged by the veterinarian. Sometimes the technician or the receptionist is the one to discharge patients. You may even be asked to assist in this process.)
- Are there specific time slots for particular types of appointments?
- How long are the time slots for appointments (i.e., 10 minutes, 15 minutes, 20 minutes, 30 minutes)?
- How many time slots are set aside for a multiple-pet family?
- How is time scheduled for house calls or farm calls?
- What about "catch up" time during the day for the doctor to write medical records, call clients, return phone calls, perform procedures on in-house patients, or other duties?
- If the veterinarian gets a call from a client, what time can you tell the caller that the doctor is likely to call back?
- Can the veterinarian be interrupted if a phone call comes in while she is with a client? What would be the circumstances?

Cross-Training

Cross-training is a simple, but powerful, technique to enhance the value and versatility of every member of the veterinary health care team. When you are cross-trained, you not only know how to do your own job, you also learn other people's duties to the level that you can perform their jobs if needed. This technique increases the flexibility of each staff member. It also increases the comfort level of staff members if one person gets tied up doing a particular

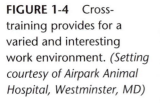

FIGURE 1-4 Cross-training provides for a varied and interesting work environment. *(Setting courtesy of Airpark Animal Hospital, Westminster, MD)*

task, is on vacation, or is called out of the office unexpectedly. Cross-training also makes your own job more interesting by providing variety (Figure 1-4).

Even if cross-training is not a regular part of the routine at the practice or hospital where you work, you may want to consider asking about being trained to perform additional duties within the practice when you have mastered your own job. It is possible that no one previously has expressed interest in being cross-trained. Most veterinary practices, however, find that having staff members qualified to perform multiple tasks increases everyone's efficiency.

OFFICE POLICIES AND PROCEDURES

Most veterinary practices have a document that provides guidance for policies and procedures at the practice. Part of this documentation may be referred to as the **personnel manual** or personnel policies. The personnel policies address issues such as attendance, dress code/appropriate attire, vacation, sick leave, codes of conduct, and others. These policies outline the employer's expectations and provide a basis of consistency in how practice issues are handled.

A **procedures manual** may be included in the personnel manual or personnel policies, or it may be a stand-alone document. The procedures manual contains information about the regular events of the practice, as well as patient policy information. Examples of the information typically covered in a procedures manual include:

- opening and closing procedures for the facility
- how patients will be admitted and discharged
- what to do in case of fire, emergency, attempted robbery, client injury, and so on.
- how the phone is to be answered (every practice has its own preference and style)
- how receipting and invoicing should be handled
- whether the practice does billing, or if all transactions are on a cash basis
- information about credit card receipting
- the basics about the practice's computer system (most practices conduct specific staff training on how to use the computer system)

Usually, you will find that the procedures manual outlines, in detail, the day-to-day workings of the practice or hospital. You should become very familiar with the practice's policies and procedures, to know where to look in the manual for clarification and to be aware of any additions or changes in the manual. The personnel policies and the procedures manual provide the foundation as you establish your own routine in the practice where you work.

SUMMARY

In addition to the veterinarian, most veterinary practices also employ veterinary technicians, veterinary assistants, practice managers, receptionists, and bookkeepers, each of whom plays a vital role in the day-to-day functioning of the office. Veterinarians and veterinary technicians are primarily health care providers, while practice managers, receptionists, and bookkeepers focus on the administrative end of the business. Veterinary assistants, however, fall somewhere in between, aiding the medical practitioners and fulfilling certain administrative duties.

Veterinary care is provided in a variety of settings, including individual and group practices, veterinary hospitals, and a variety of corporate veterinary entities. Regardless of the setting, as a member of the veterinary team you need to understand what is expected of you and be able to perform as needed. Be aware of how the practice typically does business, and be familiar with the policies and instructions in the procedures manual.

CASE STUDY

It is Ella's first day on the job as a veterinary assistant at Dr. Parker's practice. She arrived a few minutes early, hung up her coat, and then took a seat near the office manager, Trisha, while she waited for something to do.

They sat and talked about the practice for a few minutes, and then two people came in with their pets. Ella watched Trisha as she greeted the patients and took their information, then the phone began to ring. Noticing Trisha was still busy with the clients, Ella decided to answer the phone.

"Um, hello?"

"Hi, is this the veterinarian's office?"

"Yes it is, can I help you?"

"Well, I hope so. My dog, Pepper, has a runny nose, and she's been walking on her hind legs funny, as if she's in pain."

"Hmmm . . . it sounds like she's just got a cold, and she probably just bruised her leg when she was out running and playing. Dogs do that a lot. I bet she'll be fine in a few days."

"Well, I'd really like to her to see the doctor if that's possible."

"I don't know. I guess you could bring her by."

"When should we stop in?"

"I don't really know, I just started here and I'm not the receptionist. Could you call back later? Maybe in a half hour or so?"

"Okay, I'll do that. Thank you."

"No problem. Bye!"

- What did Ella do wrong?
- What did Trisha do wrong?
- What should Ella have been doing the morning of her first day on the job?
- What kind of preparation might have made this situation better?

REVIEW

Indicate whether statements 1–5 are true or false.

1. A veterinary assistant may be responsible for performing diagnostic analysis of blood samples.

2. The veterinary assistant is solely responsible for discharging patients after surgery.

3. A veterinary technician gives medication to animal patients and performs tests on them.

4. A veterinary assistant may be responsible for administering anesthesia or other medications during surgery.

5. Appropriate dress requirements for the practice in which you work will be outlined in the procedures manual.

6. List five policies of the practice that you will need to know if you will be scheduling appointments.

ON-LINE RESOURCES

Occupational Outlook Handbook

Part of the Bureau Labor Statistics Web site, the Occupational Outlook Handbook offers information on a variety of careers, including animal care and service workers.

> <http://www.bls.gov>
> Search Term: Occupational Outlook Handbook

American Veterinary Assistants Association (AVAA)

The official home page of the American Veterinary Assistants Association, this site provides information on the veterinary assistant field, message boards, and other news and information related to the field of animal care.

> <http://www.avaa.bigstep.com>

American Veterinary Medical Association (AVMA)

The official AVMA site features news articles, links to professional publications, education resources, and governmental relations, report information, a veterinary career center, and animal care information.

> <http://www.avma.org>

National Association of Veterinary Technicians in America (NAVTA)

The NAVTA Web site features information on promoting veterinary technicians, career building and credentialing information, listings of state representatives and student chapters, the veterinary technician code of ethics, and links to related sites.

> <http://www.navta.net>

American Animal Hospital Association (AAHA)

The AAHA Web site features news, an on-line bookstore, annual meeting information, on-line classified ads, continuing education information, hospital accreditation information, and information about the association itself, including a history and membership information.

> <http://www.aahanet.org>

Employer-Employee.com

This Web site focuses on issues faced in many office settings, including issues of policy and procedure, management, team building, and improving lines of communication.

> <http://www.employer-employee.com/>

CHAPTER 2

Care and Maintenance of the Veterinary Practice Facility

OBJECTIVES

When you complete this chapter, you should be able to:

- explain the importance of maintaining a clean, safe, and organized veterinary facility, and describe what steps must be taken to do so
- explain what you should do when you find that a piece of equipment or some aspect of the facility needs to be repaired
- identify common safety hazards
- identify common sources of accidental injuries and what you can do to protect yourself

- explain general techniques for avoiding infection and the spread of disease among humans and animals in the practice
- explain what OSHA is and describe some of the general guidelines that must be followed in the practice
- explain how to order and maintain office inventory and supplies

KEY TERMS

Occupational Safety and Health Administration (OSHA)	squeeze chute	centrifuges
	right-to-know station	medical supplies
clean as you go	material safety data sheets (MSDS)	pharmaceuticals
autoclave		inventory
sharps container	capital equipment	want list

INTRODUCTION

Overseeing the general maintenance of a veterinary practice can be a full-time job, and many details must be addressed. The cleanliness of the practice facility is critical, not only to ensure delivery of high-quality veterinary care to animal patients, but also to communicate to clients the quality of service delivered at the practice. Plans must be in place to deal with the unexpected breakdown of an essential piece of equipment, whether medical (such as an anesthesia machine), or nonmedical (such as a furnace or air conditioner).

All veterinary practices are required to comply with the Occupational Safety and Health Act of 1970, regulated by the **Occupational Safety and Health Administration** (OSHA), which provides job safety and health protection for workers by promoting safe and healthful working conditions (Figure 2-1).

Most veterinary practices have identified specific strategies to help their employees prevent injuries. You may be required to read and comprehend a safety manual, in addition to the personnel manual and the procedures manual. There are also safety precautions that must be taken to help minimize hazards to clients visiting the practice. Preventing the spread of disease within a veterinary facility from one animal patient to another is an important aspect of hospital care and maintenance.

FIGURE 2-1 OSHA guidelines that provide for a safe work environment must be posted in the practice where they can be seen easily by all staff. *(Photograph courtesy of Robin Downing, D.V.M.)*

MAINTAINING A SAFE FACILITY

Housekeeping and General Cleaning

If it smells clean, then it must be clean.

These words address one of the most important aspects of the general care and maintenance of the veterinary practice: odor control. It is impossible to achieve adequate odor control in any veterinary facility without making cleanliness a priority. A clinic can look clean, but if it smells dirty ("like animals") then the issue of general cleanliness is not receiving enough attention.

Many veterinary practices use a professional cleaning service to take care of the general cleaning of the physical facility on a regular basis. The schedule of a professional cleaning service will vary with the needs of the practice. A high-volume, high-traffic practice may have that cleaning done every night after the clinic is closed. Another practice may have professional cleaning done weekly and have the staff keep up with things during the rest of the week. Some practices use only their own staff members to keep the facility clean. Whether or not your office uses a cleaning service, you should be sure to pick up after yourself and tend to any accidental messes that pose a health risk (i.e., pets urinating in the waiting room).

It is everyone's responsibility to help keep the practice clean on a daily basis. You will need to keep an open mind about what may be required of your position within the practice where you work. For many veterinary assistants, the job description includes cleaning cages and feeding, watering, exercising, and bathing animals, all of which relate to keeping the practice clean. You will need to learn what cleaning products are used in the different areas of the facility. A chemical that works very well on the floor and cuts grime where people have walked might actually cause harm to an animal if it were used inside a cage, a stall, or a dog run. If you are not sure whether a product is safe to use for a particular purpose, ask! Industrial cleaners can have very harmful effects on humans and animals if used improperly.

Take pride in the facility where you work, and contribute wherever you can to the overall housekeeping of the practice. As a veterinary assistant, you will probably have the opportunity to participate in activities and procedures in many different areas of the practice. Most cleanliness issues involve common sense. Adopt an attitude of "**clean as you go.**" This simply means putting things back where you found them when you are finished using them, tidying up an area after a procedure, clearing counters that have become unnecessarily cluttered, and dusting (Figure 2-2). This might also include wiping down a work area with an approved disinfectant, sterilizing instruments with an **autoclave**, and disposing of needles in a **sharps container**. Be sure to ask your supervisor about specific cleaning procedures and products.

Most veterinary practices that serve companion animals want to minimize cleanup of animal waste—the most common source of spreading disease among animals. One way to accomplish this goal is to recognize that dogs need

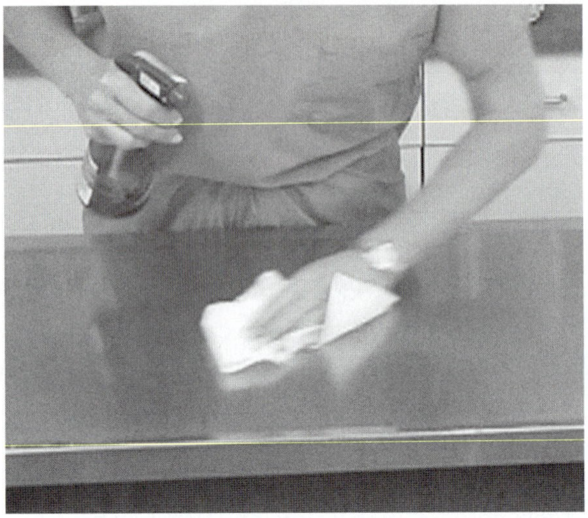

FIGURE 2-2 Proper cleaning is essential to the maintenance of the veterinary facility.*(Setting courtesy of Airpark Animal Hospital, Westminster, MD)*

to be walked outside in an exercise area on a regular time schedule so that they may empty their bowels and bladders. Remember, dogs do not like to soil where they live and sleep. They have been housetrained and recognize inside soiling as unacceptable. Be aware of their comfort by providing them with adequate exercise/elimination opportunities. You will enjoy the additional benefits of decreased cleanup, decreased risk of offensive odors, and, most importantly, decreased risk of disease within the veterinary facility. (However, it is important to note that the exterior exercise and elimination area must also be kept clean.)

When caring for cats, keeping litter pans clean is the most important step in preventing undesirable odors and diseases. Most veterinary facilities have a morning and evening routine that involves scooping cat litter pans (Figure 2-3), changing water, and feeding.

If canned food is fed to cats, be sure to remove any uneaten portions (and note which animal(s) did not eat). Old canned cat food can emit quite an unpleasant aroma! Cleanliness also directly affects cats, which are very clean and particular animals. They do not appreciate the accumulation of waste in their litter pans. If a cat is in the veterinary hospital because it is ill, you want to do all you can to encourage a healing attitude. Keeping a cat's cage clean is a good beginning.

Within a veterinary facility that serves large animals, general cleanliness and odor management guidelines are just as important as in a small-animal practice. For instance, if you have been working with the veterinarian on cattle in the morning and have become soiled with manure, leave your boots and coveralls at the door (and clean them with an appropriate disinfectant as soon as possible). If the veterinary hospital has stalls that house large animals, be aware of your role in keeping the stalls cleared of manure- and urine-soiled hay. Just look around you throughout the day and ask yourself, "If my animal were staying here, would I be happy with the environment?" This attitude is a good way to stay in touch with how comfortable and clean the animals are at any time.

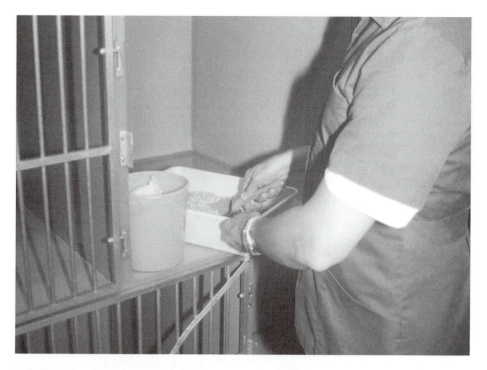

FIGURE 2-3   Keeping cages and litter pans clean is part of the job in a small-animal practice. *(Photograph courtesy of Robin Downing, D.V.M.)*

Safety Hazards

In any veterinary practice, the potential is great to encounter a wide variety of safety hazards. Many of these hazards are unique to the veterinary profession. Animals may bite, scratch, or kick. There are needles, scalpel blades, and other sharp objects to avoid and dispose of properly. Chemicals, anesthetics, animal urine and feces, and the discharge from infected wounds can pose a health risk to humans if not handled properly. Within the veterinary facility, there may also be equipment that can cause injury.

In practices that serve large animals, heavy gates and spring-loaded latches can pinch fingers, as well as catch arms and legs (Figure 2-4). The radiation area of the hospital demands respect, and you will need to learn how to prevent exposure to x rays. Lifting animals, bags of pet food, and containers of animal bedding material can all pose potential problems. It is very important to know how to keep yourself as safe as possible.

Some of the medications used in veterinary practice can also pose a variety of health risks. Certain vaccines used with large animals can cause disease in a person who is accidentally inoculated. Anesthetics are drugs that must be treated cautiously. If the practice where you work treats cancer patients, you will need to learn how to minimize your exposure to potent chemotherapy drugs. Anesthetic and chemotherapeutic drugs are quite dangerous because of the harm they may cause if inhaled. Some antibiotics should be handled carefully as well.

FIGURE 2-4   Machines with heavy gates used to restrain large animals can pose a threat to veterinary workers.

Preventing Accidental Injury

The most common injuries in veterinary practices are bite wounds from animals. Dogs and cats usually come to mind first, but do not forget that horses, ferrets, rabbits, and reptiles can all inflict serious bite wounds as well. Most bite injuries are preventable if animals are handled sensibly. If you do not have much experience handling animals in a formal setting like a veterinary practice, you must take the initiative to seek guidance and assistance. Handling animals safely takes practice.

You should begin your training by handling those animals that are not stressed by being in a veterinary hospital. You will need to handle them in a variety of settings such as getting them out of their cages, putting them back into their cages, restraining them for their physical examinations, taking them outside to eliminate, and moving them from one pen to another. Prepare yourself by assuming that the animal may bite. It is also important to learn the behaviors, postures, vocalizations, and facial expressions that animals use to

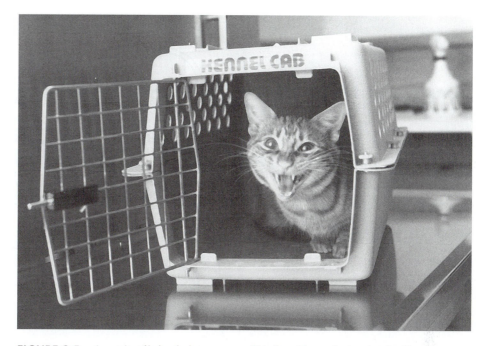

FIGURE 2-5   An animal's body language will tell you how that animal will react to you and whether you should handle the animal alone.

communicate their attitudes (Figure 2-5). You can then begin to interpret when an animal will allow you to do what needs to be done without a struggle, and when it would be best to call for an extra hand.

The most common piece of restraint equipment is the leash. Most of the animals encountered in a standard veterinary practice are companion animals, primarily cats and dogs. A securely attached leash held short (so that it allows enough room for the animal to maneuver on the exam table, but not so much that the animal can get up and walk or jump. while being examined) exerts a great deal of control over an animal. Another useful technique is to wrap one of your arms across the front of the animal, to keep it from lunging forward, and another across its back, to hold it in a sitting position and to dissuade it from moving. If the need arises, you can quickly join your hands and tighten your grip to hold the animal. Please note that this technique may not always be appropriate, especially if the animal is visibly agitated, known to be violent, or physically larger and stronger than a person of your stature and strength can reasonably handle. If you are unsure of how to restrain a particular animal, speak to the veterinarian before the examination begins.

It is critical that clients are not allowed to handle or restrain their animals while the animals are under the care of the veterinary practice. If a client is injured in that circumstance, the veterinary practice may be held legally liable. As tempting as it may be to allow an animal owner to hold and reassure his or her frightened pet, it is simply not worth the risk of the client being injured.

Besides animal bites, other accidental injuries common to a veterinary practice usually involve some aspect of the physical facility and tend to be

FIGURE 2-6   You should learn the proper way to use a squeeze chute to avoid injury.

avoidable with a bit of care and common sense. It should go without saying that if your practice has a **squeeze chute** (a small stall used to restrain large animals), as shown in Figure 2-6, you should be specifically trained in how to safely use that chute, which can easily cut off a finger. It is also wise to remember that large animals have a physical advantage, so you must be aware of the nearest escape route when handling them. And remember to avoid working with large animals when you are alone.

Back injuries due to lifting—very common in veterinary practices—can be avoided, in many cases, through a bit of planning. Remember that large dogs usually are not used to being lifted. They may struggle and throw you off balance. It is best to lift them with the help of another, and always lift with your legs, not your back. Your employer probably will ask you to follow weight guidelines that suggest a maximum weight you should attempt to lift alone (whether you feel you could lift more than that amount or not). There may be protective equipment, such as back-support belts, for you to wear when you are lifting heavy animals or inanimate objects. Some practices have a power-driven table to use for lifting large, heavy animals.

During cleanup procedures, be careful of slippery floors. Remain alert for a needle that has not been placed into the appropriate sharps container. These containers are made of heavy plastic with a lid that can be permanently sealed. When the container is full, the lid is sealed and the container is picked up by a service licensed to dispose of hazardous waste materials. When you find an unlabeled container, bring it to the veterinarian's attention immediately. By remaining alert to your surroundings, you usually can avoid circumstances that might otherwise lead to accidental injuries.

Controlling Infection and Spread of Disease

It is especially important, when working in a veterinary facility, to be aware of the risks of passing germs, bacteria, viruses, and diseases from one animal to another. Most animals are admitted to a veterinary hospital because they are sick. Their immune systems are not working at full capacity from the stress of their illnesses, so they are more susceptible to disease. The last thing a veterinarian wants is one sick animal serving as the source of illness for another. Remember that animals in the hospital for surgical procedures experience more stress than usual—making them more likely to pick up germs if the veterinary staff is not careful.

For disease containment, like many other aspects of veterinary medicine, common sense considerations will often help you make wise choices when you are handling patients. Animals suffering from infectious diseases should be kept in isolation to avoid infecting other patients. Ill animals should have their daily maintenance, like feeding and watering, done after the healthy animals to decrease the risk of carrying germs from the sick animals to the well ones. Wash your hands between handling all animals. Change into a clean smock whenever you have handled an ill animal or if the smock has been soiled with urine, stool, blood, pus, or other body fluids. Never wear the same smock for feeding, watering, or medicating animals that you used for cleaning their cages. A separate smock should also be used for bathing animals.

You must learn and be comfortable with all procedures outlined by your veterinarian for reducing the risks of transmitting disease from one animal patient to another. If you come upon a situation with which you are not familiar, or if you are unsure how to proceed, it is always best to seek out a senior employee and ask for guidance. There may be circumstances when a truly infectious case is hospitalized in the facility's isolation ward. You may or may not be authorized to handle that particular case. Every practice has its own requirements, and it is your responsibility to do all you can to keep the patients of the practice safe from unnecessary risk.

OSHA GUIDELINES

All employers are required to abide by the Occupational Safety and Health Act of 1970. Commonly referred to as OSHA, it provides for safe and healthful working conditions for all employees. The guidelines were created by the U.S. Department of Labor. The employer is required to inform all employees of potential hazards in the workplace so that the employees may understand how best to keep themselves safe. OSHA regulates many aspects of businesses in all industries in the United States. Examples of safety regulations that must be adhered to include the following, which require veterinary practices and hospitals to:

- keep the location clean, safe, and dry
- provide easy access to clearly marked, adequately-sized, unobstructed exits

- ensure that compressed gas cylinders (such as those that might be used for anesthetic purposes, for example) are properly marked and used
- follow strict safety guidelines when handling hazardous chemicals
- require the use of proper protection for eyes, face, skin, feet, respiratory, hands, and so on, in situations posing an environmental, chemical, situational, radiological, mechanical, or any other form of physical hazard
- provide toilet facilities and clean water
- maintain operating fire extinguishers, sprinkler systems, and other fire extinguishing systems (requirements can vary by location, type of business, materials involved, etc.)
- include protective guards or covers for machinery
- keep electrical equipment free of hazards
- provide information to employees regarding all potential hazards of using equipment, hazardous materials, chemicals, and the like.

Keep in mind that these are only a small sampling of some general OSHA rules. Your employer will have more information about the specific regulations that must be followed in the practice or hospital. Additional information about OSHA regulations are available on-line at <http://www.osha.gov>. OSHA can also be reached at 1-800-321-OSHA (6742), or via mail at U.S. Department of Labor, Occupational Safety & Health Administration, 200 Constitution Avenue, Washington, D.C. 20210.

Right-to-Know Station

Somewhere in the practice facility will be a **right-to-know station** (Figure 2-7), where there is a notebook or binder containing the safety information that applies to that particular practice. Contained in the safety notebook will be **material safety data sheets** (MSDS), which are published by the manufacturers of the various products used by the practice. Read this information to be aware of any special handling requirements of the products or any restrictions that may apply to the user of the products. This information varies for each product used by the practice.

The safety notebook will also contain (1) evacuation plans in the event of an emergency; (2) the locations of gas and water valves for rapid shutoff (if required); (3) the locations of fire extinguishers (if applicable); and (4) emergency telephone numbers for the police, the fire department, and companies that service the facility's heating, ventilation, and air-conditioning units. All chemicals, disinfectants, and other products that are used within the practice will be labeled by your employer with any appropriate hazardous considerations. In addition, personal safety equipment like safety glasses, latex gloves, and ear plugs will be supplied by your employer.

Your employer, the veterinarian, or the designated safety officer within the practice will show you where the safety information and safety equipment are kept. It is your responsibility to read and understand the practice's safety manual, to know how to use the required personal safety equipment, and then

FIGURE 2-7   Example right-to-know station. *(Photograph courtesy of Robin Downing, D.V.M.)*

to actually use it. It is also your responsibility to be comfortable reading MSDS for appropriate precautions. You must take the initiative to remain alert to potential hazards you encounter within the practice facility and then bring them to the attention of the appropriate individual. Safety is everyone's business.

In order to avoid losses from preventable injuries, most veterinary employers have developed strategies to help their workers anticipate potentially hazardous situations. Usually, they will supply you with a document that outlines the steps to take in the event that you witness or are involved in an injury at the practice. If you experience or witness an accident in which a staff member, client, or visitor sustains a personal injury, regardless of how serious, immediately report the situation to your supervisor or to the person in the practice who oversees safety issues. Failure to report an accident or injury can result in a violation of legal requirements and could delay the processing of appropriate insurance claims and benefits.

OFFICE EQUIPMENT AND MAINTENANCE

Repairs and Troubleshooting

Part of helping a veterinary facility run smoothly involves keeping the equipment, both medical and nonmedical, in good working order. As a veterinary assistant, you will be using and handling many different pieces of equipment as you complete your duties. It is unlikely that you will be called upon to repair any of that equipment, but it is your responsibility to learn the appropriate way to use any equipment or tools that are the property of the practice. If there is no formal training program in place to prepare you to use the equipment in the practice, seek out a senior employee who is able to demonstrate correct usage. You may also be asked to participate in regular, routine maintenance of the practice's equipment. Proper use and maintenance might include:

- a daily or weekly cleaning or calibration routine
- replacing paper in printers, photocopiers, and fax machines
- replacing ink or toner cartridges in printers, photocopiers, and fax machines
- learning how to properly turn each piece of equipment on and off
- learning how to operate machinery and equipment you are not familiar with
- learning how to take readings on various pieces of veterinary and laboratory equipment
- learning how to identify technical problems

Most veterinary practices are now computerized, so you will probably be working with a computer station and keyboard as part of your job (Figure 2-8). Although you are probably at least somewhat familiar with computers, you may not be skilled in using the specific veterinary programs used in the office. Be sure to request training in the use of the computer so you can use it both

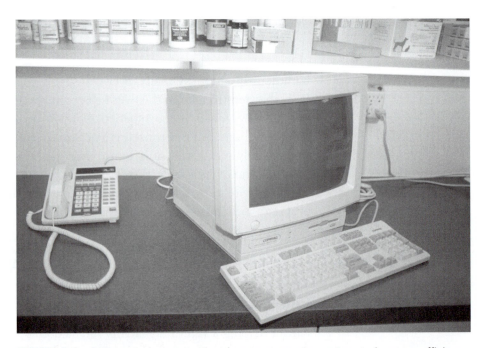

FIGURE 2-8 Most veterinary practices have a computer system to increase efficiency in the front office. *(Photograph courtesy of Robin Downing, D.V.M.)*

correctly and effectively. (Computers and software are discussed in greater detail in Chapter 4.)

In addition to computers, you can expect to find and use many other (sometimes complex) pieces of office equipment. There may be a typewriter for typing medical record labels. Sophisticated telephone systems for the veterinary practice allow multiple phone lines to come into a single reception area, and many systems have built-in voice mail and paging capacities. The practice where you work may use a telephone answering machine after hours. You may be asked to record the outgoing message and get the machine ready to receive calls each day at the end of business hours. A fax machine, now common in veterinary offices, may be a stand-alone piece of equipment or may also include a phone (Figure 2-9). Your practice might also use a multifunction machine, incorporating a printer, fax machine, copier, and/or scanner.

You may be asked to become familiar with the use of an adding machine (Figure 2-10), a photocopier, a cash register, a postage scale, and a postage meter. The practice may use audiovisual equipment for client education or staff-training purposes. There may be a TV/VCR for use at the facility. Some practices use endless-loop videotapes that play the same piece over and over. No matter what equipment you find in the practice where you work, part of your responsibility is to become very familiar with how to use, care for, and manage it—and whom to contact if something unusual happens.

In addition to learning how to use various pieces of equipment to their best advantage, you need to learn to perform basic maintenance and troubleshooting. For instance, you should know how to load paper into the printers attached to the computers. The process is different if the printer is a bubble jet, laser, or dot-matrix type. Paper tapes in adding machines must also be replaced when they run out.

You may be asked to change the ribbon in the typewriters or the cartridge in a computer printer. If the practice has a photocopier, you may occasionally

FIGURE 2-9   A fax machine with phone attached is a common piece of equipment in a veterinary practice. *(Photograph courtesy of Robin Downing, D.V.M.)*

FIGURE 2-10 You may be asked to become familiar with using an adding machine. *(Photograph courtesy of Robin Downing, D.V.M.)*

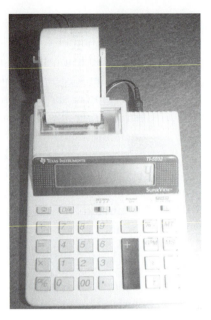

be asked to replenish toner or empty a full toner receptacle (Figure 2-11). You will want to become familiar with refilling the paper trays in the copy machine and the printer, as well as knowing where replacement paper is stored.

Although you may not be required to repair equipment in the practice, you will need to know the person in the practice to contact if there is a malfunction. Most likely, you will report a problem with front-office equipment, like the photocopier or the fax machine, to the office manager or receptionist. In the treatment and surgical areas of the practice, a senior veterinary technician may be responsible for the equipment. It is possible that one of the veterinarians in the practice oversees the medical equipment. Be aware of

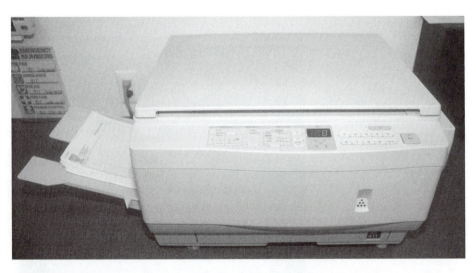

FIGURE 2-11 You should become familiar with using the photocopier, as well as replacing the paper and changing the toner. *(Photograph courtesy of Robin Downing, D.V.M.)*

the optimal temperature in different areas of the practice facility so that you will notice any sudden changes. Be alert to hear the furnace making strange noises or an exhaust fan failing.

Timing is usually important in a medical practice, so you will want to report any abnormalities that you notice immediately. Do not assume that the appropriate person already knows of the problem. To avoid causing further damage, do not use any equipment that is functioning abnormally or is sending a message that a part is not working. By ignoring a minor problem, you may allow it to become a major one. You may be asked to contact the designated repair or service person, so you will want to be familiar with where a list of those support people is kept or who in the practice has access to that information.

MATERIALS AND SUPPLIES

Veterinary practice facilities use many different types of materials and supplies. You would expect to find some of these in any office setting. Others, very specific to the veterinary profession, require detailed record-keeping techniques.

The supplies that help office equipment operate will likely be found in the administrative area of the practice. Most often, the lead receptionist or the practice manager is in charge of keeping an acceptable inventory of those supplies. Supplies for the administrative area of the practice include copy paper, computer paper, printer cartridges (ink or toner), typewriter ribbons, adding machine tapes, pens, pencils, markers, telephone message books, and anything else that is necessary for a smoothly functioning office. You will need to be familiar with the storage area for various administrative office supplies in case you are filling in at the reception desk and find, for example, that the copy machine is out of toner. You should also know exactly who to talk to when you need to order more, and the requisition process, if applicable.

Another category of veterinary practice supplies includes the **capital equipment**—equipment that has a fairly long life expectancy and contributes to the income of the practice. Examples of capital equipment include in-house laboratory equipment like blood analyzers, **centrifuges** (for separating blood components), and microscopes (Figure 2-12). Other capital equipment in a typical veterinary practice includes an anesthesia machine, an x-ray machine, equipment for performing dental procedures, and an autoclave (for sterilizing surgical instruments).

A third category of supplies necessary for the smooth functioning of a veterinary practice is **medical supplies**. This category includes items like bandaging and suture materials, intravenous fluids, needles, syringes, x-ray film, chemicals that are used in the laboratory, and others. These items, used in the everyday functioning of the practice, help the veterinarians and veterinary technicians do their jobs. Equally as important as medical supplies are the **pharmaceuticals** used in the practice facility, as well as those dispensed to clients from the practice pharmacy. You would expect to find a variety of vaccines for various species of animals and many different diseases among the

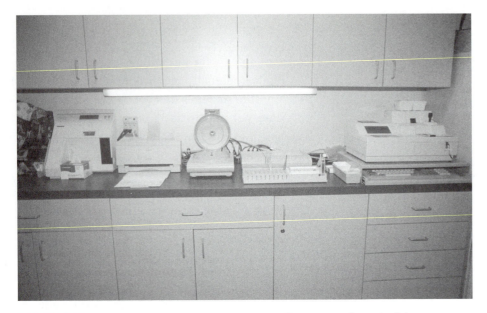

FIGURE 2-12 Some veterinary practices have a fully equipped on-site laboratory. *(Photograph courtesy of Robin Downing, D.V.M.)*

pharmaceuticals in a typical veterinary practice. You will also find miscellaneous medicines including antibiotics, anesthetics, sedatives, medicated shampoos, ointments, deworming medications, vitamins, and minerals.

Want Lists

Few things are more frustrating in any work situation than to be in the middle of a project or procedure and suddenly run out of needed supplies to finish the task at hand. In a medical setting, this scenario can have deadly implications. It is therefore critical for all staff members to know the procedure used to avoid shortages of practice **inventory**. Most practices have very specific staff expectations concerning keeping appropriate levels of products and supplies on hand. However, several common sense concepts have universal application.

One very useful tool is the **want list**—a list posted in a notebook, on a clipboard, or on a white board at a specific location in the practice for all to see and use. When you retrieve a particular medication or supply from the practice inventory and notice that the remaining quantity has dropped to a low level, it is time to enter that item on the want list (Figure 2-13). Most practices have a minimum quantity of commonly used products that is to remain on hand. At regular intervals, the person in charge of ordering practice materials retrieves the want list and uses it to guide the purchasing for a particular time period—whether ordering is done on a weekly, monthly, or quarterly basis.

Some practices use separate want lists for medical supplies and for office supplies. Breaking it down further, you may find a want list for the front-office

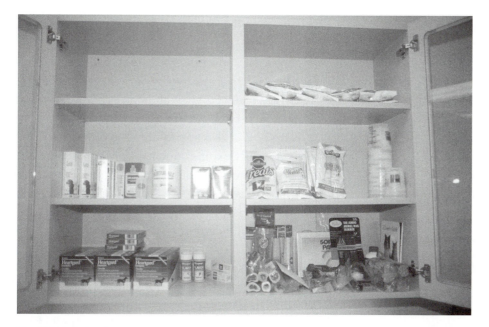

FIGURE 2-13   When items for use or sale by a practice fall below a certain quantity, that information should be recorded on the appropriate want list. *(Photograph courtesy of Robin Downing, D.V.M.*

area, another for the pharmacy, and yet another for the treatment/surgery area of the practice.

Inventory Management

Most practices and hospitals use computer systems to assist with purchasing decisions. If the inventory for the practice is entered into the computer system and tied into the invoicing function of the practice program, any inventory items entered on an invoice will automatically be removed from the practice inventory. Periodically, the computer program can then be accessed to generate a report suggesting necessary purchases.

Usually, a veterinary practice has a single person responsible for placing orders for inventory items for the practice, whether or not a computer is used in the purchasing process. That person may be the office manager, the practice manager, the bookkeeper, or the practice owner. A one-person system helps avoid duplication. It also avoids confusing someone in the practice who may think that another had placed an order—leaving the order undone. One of your jobs, as a veterinary assistant, will be to record, in a timely fashion on the appropriate want list, any items that you notice are running low. You also may be asked to report shortages directly to the person involved with ordering.

Although you may not be involved directly in ordering, it is highly likely that you will be involved, in some way, in processing orders when they arrive. When an order is received, the first step is to be sure that there has been no damage to the package. If a shipment arrives damaged, you should follow whatever guidelines are appropriate for the particular carrier or delivery system. Step two is to open the package, find the packing slip or invoice, and carefully examine the contents to be sure that all items listed are present. This should be done prior to signing for the delivery, as your signature indicates that the shipment was received, with all appropriate contents and in acceptable condition. Report any shortages in a timely fashion to the appropriate staff member. At this time, it is also appropriate for you to find and record on the invoice any listed product expiration dates. This procedure allows the person entering information about the order in the practice's computer to include these dates. It also provides a means of screening shipped products that have been *short-dated*—that is, products that will not be used up before their expiration date. This procedure gives the practice owner, or whoever handles inventory management, the opportunity to contact the company about an exchange.

When you are confident that everything on the invoice is present and you have recorded the expiration dates of products when appropriate, put everything in its correct location in the practice facility. Certain drugs and lab tests require refrigeration or freezing. These should be checked in immediately and then refrigerated as necessary before moving on to the remainder of the delivery. Remember to *rotate* the inventory as you put things away. Rotating the inventory—moving products that are already on the shelves to the front and placing newer items behind—ensures that products are used by the practice or sold to the clients in the order in which they were received. Rotation cuts down on the risks that products will become outdated before they can be used or sold.

Occasionally, even with the most accurate computerized inventory records, you will be asked to assist with a *physical inventory*. This means counting every single inventory item in the practice and making a notation of the quantity on an inventory list. You may find that the list for office supplies is kept separately from the medical supplies list. The pharmaceuticals may be on their own list as well. Many practices conduct its physical inventory at the end of its fiscal year, which may not be the same as the calendar year. Some practices perform a physical inventory at the end of each quarter in an effort to have the inventory quantities in the computer remain as accurate as possible. As a veterinary assistant, you need to keep an open mind about your role in inventory management, because it can vary dramatically from practice to practice.

SUMMARY

Cleanliness, safety, the containment of infectious disease, and the general maintenance of a healthful facility are the responsibility of all persons involved in a veterinary practice. Each member of the veterinary staff must be involved in identifying and correcting hazards and potential contamination, and as a veterinary assistant, you are especially responsible for maintaining sanitary conditions for the animals in your care.

To maintain a safe working environment, all members of the veterinary staff should comply with OSHA regulations and use general common sense in daily activities. If something might be too heavy to lift alone, ask someone to help. If a piece of machinery poses a hazard, be sure to use care and caution while operating it, and do so only after being adequately trained.

Office equipment, materials, and supplies are also important to the smooth functioning of a veterinary practice, and the use of want lists helps with inventory management, ensuring that when something is needed, it is available and in optimal condition. The overall care of the veterinary facility is essential to the delivery of excellent medical services.

CASE STUDY

On Wednesday morning, Janet accompanied Dr. Eggers on a farm call to help him with inoculating some horses. When they got back to the office, Janet entered through a rear door, leaving her boots and coveralls outside, away from the animal walking area. As she entered the reception area, Lisa, the receptionist, mentioned that the lab had just called to let them know they were faxing over the results of a patient's blood work.

Janet noticed that there was no fax coming through, then checked the machine and realized that it was out of toner. Going to the supply closet, she picked out a new cartridge and put it into the machine so the fax would print. Then she boxed up the old one to be recycled and asked Lisa to add a fax toner cartridge to the front office want list.

Next, she went down the line of cages in the back, feeding each animal and clearing out waste. Peanut, Mrs. White's dog, quivered and raised his lip when Janet opened his cage. Leaning in, she patted him on the head and he snapped at her, scratching her hand with his teeth. Janet quickly closed his cage and went to clean the wound. Thankfully, the cut was not deep, but to be safe, she checked with Dr. Eggers to see if there was anything she should be concerned about.

- How did Janet act proactively to create and maintain a clean, safe, and functional work environment?
- Which of Janet's actions might have helped to spread germs among the animals in her care?
- What preventive measures could Janet have taken to avoid the dog bite?

REVIEW

1. *True or False?* Uneaten portions of canned cat food should be left in the cat's cage until the cat finishes the food.

2. What are some significant steps that you can take to keep the facility clean?

3. What should you do if you notice that an exhaust fan in a kennel area is making an unusual noise?

4. List five potential safety hazards for employees of a typical veterinary practice.

5. List three precautions you can take to minimize the risk of injury by animals.

6. List five OSHA requirements for ensuring safety in the veterinary practice.

7. What can you learn from material safety data sheets (MSDS)?

8. List four types of information, other than MSDS, included in the safety notebook housed in a practice's right-to-know station.

9. List eight types of equipment typically found in the office of a veterinary practice.

10. List five examples of capital equipment used by veterinary practices.

11. What purpose does rotating the inventory serve?

12. List five examples of pharmaceuticals used by veterinary practices.

13. Explain the significance of a want list.

14. What are the steps you should take in processing orders when they arrive?

ONLINE RESOURCES

Veterinary Information Network (VIN)

VIN, an on-line service for veterinary professionals, offers a reference center, OSHA information, and classifieds for the veterinary industry. Membership is free for students at sponsored universities.

<http://www.vin.com>

Occupational Safety & Health Administration (OSHA)

Part of the U.S. Department of Labor, OSHA's mission is to prevent injury, illness, and death in the workplace. The Web site offers information on safety regulations, programs, services, and other resources related to workplace safety.

<http://www.osha.gov>

VetGuide.com

VetGuide is the illustrated buyer's guide to new equipment and supplies for the veterinary facility, offering links to a wide variety of veterinary products for both the doctor and the front veterinary office.

<http://www.vetguide.com>

Animal Care Equipment & Services (ACES)

This site features numerous types of animal care and handling equipment, and includes images of various cages, protective devices, safety vests, and veterinary accessories that a veterinary assistant might encounter in a veterinary office.

<http://www.animal-care.com>

CHAPTER 3

Administrative Duties

When you complete this chapter, you should be able to:

- identify the types of records found in a typical veterinary practice, including laboratory reports, radiographs, and medical records
- describe the maintenance of the various logs found within a typical veterinary practice, including those for surgery/anesthesia, laboratory tests, radiographs, and controlled substances
- describe how to admit and discharge patients,

take patient histories, maintain records, and prepare appropriate release forms and certificates for signature

- describe basic concepts of records retrieval and protection of medical and business records
- identify basic filing systems and filing equipment
- explain how to screen and process mail efficiently

medical records	SOAP	alphabetic filing
objective information	purge	admitted
subjective data	logbook	consent form
source-oriented medical record	x-ray logbook	discharge
problem-oriented medical record	anesthesia/surgery logbook	neuter/spay certificate
conventional format	controlled substance logbook	vaccination certificate
master problem list	laboratory logbook	health certificate
	numerical filing	addressee

INTRODUCTION

Although the specific duties of the veterinary assistant will vary slightly from practice to practice, you will undoubtedly be called upon to perform at least some basic administrative functions, such as maintaining and filing medical records and other forms of documentation, admitting and discharging patients, and screening and processing mail.

Some of these administrative responsibilities may seem a bit mundane on the surface, but they are, nevertheless, vital functions that ensure the successful operation of the veterinary facility. Medical records summarize the patient's history for the veterinarian and provide specific facts about symptoms, medications, and past treatments. Without these, the doctor would be working 'blind.' Proper admission and discharge practices help to ensure that all patients are seen in a timely manner and are properly tended to before leaving. Mail screening and processing facilitates the passage of important communications to the proper parties as quickly and efficiently as possible. As you can see, without efficient administrative practices, a veterinary facility would be unable to function.

RECORD KEEPING

Purpose of Medical Records

Medical records are the heart and soul of every veterinary practice, whether the practice sees companion animals, horses, cattle, swine, or a combination of all of these animals. They are dynamic documents that record the sequence of events each time an animal is seen by the veterinarian. Medical records are also legal documents that contain pertinent information on each animal patient, as well as on each client, and should be treated in the same manner as the medical records of a human being.

The primary purpose of veterinary medical records is to serve as a detailed description of medical issues, their progress, and their resolution in animal patients. Each record is a benchmark for measuring improvement or deterioration of an animal that has a medical problem.

The medical record provides for continuity of care by reminding a veterinarian what she has seen on the physical evaluation of a patient, and of the patient's history, the tests that have been done and their results, the diagnosis, the treatment that has been initiated, and all comments and recommendations made to the client.

A medical record also provides continuity when multiple doctors are involved in a single patient's care. It allows any other veterinarian to understand, simply by reading it, what the animal's medical history is, what the current problems are, and what is planned for the patient. Another way to think of the medical record, in this context, is as a summary of an animal's life from the perspective of its health, wellness, or chronic disease.

Another important purpose of medical records is to facilitate rapid retrieval of information about patients. For example, the medical record allows a veterinarian to review rapidly an animal's major medical problems and its medical history. A veterinarian can scan easily for episodes of illness, review the diagnostic and treatment plans, and learn the outcome of therapy. The reader can see, at a glance, all the wellness issues that have been addressed, including vaccinations, heartworm tests, surgery, lab work, and any other problems. (Figure 3-1).

The medical record allows for rapid retrieval of information about a client as well: address and telephone number, information about where the client works, and how she may be reached during the day. Quick access is especially important when an animal is in the hospital and a decision must be made in a timely fashion about a particular treatment or procedure.

A third purpose of medical records is to provide documentation of all medical decision making from a legal perspective. (Remember that medical records are legal documents.) You should record all pertinent information completely, accurately, and in a timely manner. In the event of any dispute, medical records allow for clarification of what was done and why. They protect a veterinarian and staff by providing thorough written records.

Because of the legal nature of medical records, all information should be written legibly and in ink—never in pencil. If a mistake is made, a single line should be drawn through a mistake and initialed (this indicates that an error was made rather than suggesting that information was changed). Then, the correct information should be entered and initialed next to the mistake. Do not erase, scratch out, or blot out mistakes. Medical records should be maintained continually so that they may be reviewed at any time.

Yet another purpose of medical records is to provide the means to track certain data within the patient population. Veterinarians can collect statistical information about certain illnesses. For instance, how many cases of canine heartworm were detected in the practice within a certain time period? How many limb fractures of a specific type have been seen by the practice? A veterinarian can also gather information about the makeup of his or her overall patient base by utilizing medical records. It is possible to determine how many dogs of a particular breed or how many cats of a particular age are in the patient pool.

Creating Medical Records

Each medical record must contain a certain minimum amount of information. For instance, the record should include the client's name, address, and telephone number. In addition, the client's employer and work number, as well as a number at which the client may be reached during the day, are useful. Some practices include information about the client's children, if any, or about other family members who tend to come in contact with the animal. Pertinent patient information includes the animal's name, species, breed, age, gender, reproductive status (neutered vs. intact), color, and hair coat (if applicable). Figure 3-2 shows a typical patient/client information form.

CITY HOSPITAL ANY STREET ANYTOWN, SS 00000			NUMBER

REFERRING VETERINARIAN:

Date:	NAME _____ STREET _____ CITY, STATE, ZIP _____ PHONE - BUS. _____ HOME _____
Appt. Time:	SPECIES BREED SEX ANIMAL'S NAME DATE OF BIRTH
Admitting Clinician:	COLOR—IDENTIFYING MARKS **CHIEF COMPLAINT:** _____

ENVIRONMENTAL HISTORY:
Length of time owned: **Kept In:**   **Obtained From:** ☐ Bred ☐ Breeder ☐ Friend
                                    ☐ Your State   ☐ Pet Shop ☐ Humane Society
                                    ☐ Other _____   ☐ Stray ☐ Other

ENVIRONMENT:	CONFINED TO:	OTHER PETS:	DIET:	FREQUENCY/DAY
Urban House ☐ Apartment ☐ Suburban ☐ Rural ☐	Home ☐ Outdoor pen/ chain ☐ Roams ☐ Other ☐	Yes ☐ No ☐ _____ _____ _____	Commercial Dry ☐ Semimoist ☐ Canned ☐ Table Scraps ☐ Other ☐	Amount:

PREVENTATIVE MEDICINE PROGRAM:

YES	NO		DATE	YES	NO		DATE
☐	☐	Distemper-Hepatitis	_____	☐	☐	Fecal Check	_____
☐	☐	Feline Panleukopenia	_____				
☐	☐	FVR-Calici-Panleukopenia	_____	☐	☐	Heartworm Check	_____
☐	☐	Distemper-Hep-Parainfluenza	_____	☐	☐	Heartworm Preventative	_____
				☐	☐	Flea Control	_____
☐	☐	Rabies	_____	☐	☐	Tick Control	_____

MEDICAL HISTORY:
Past Medical History—prior illness, surgery, drug reactions, etc.
Current Medical History—signs, chronological course, prior therapy, systems review.

SYSTEM REVIEW
Attitude   _____
Exercise Tolerance _____
Ocular or   _____
   Nasal DIscharge _____
Sneezing   _____
Coughing   _____
Appetite   _____
Vomiting   _____
Diarrhea   _____
P/D - P/U   _____
Pruritus   _____
Incoordination   _____
Paresis   _____
Seizures   _____
Estrus   _____

COMPREHENSIVE HISTORY   MEDICAL RECORDS

FIGURE 3-1 Properly kept medical records can quickly show the veterinarian an animal's history at a glance. Note: P/D = polydipsia and P/U = polyuria

DATE _____   CASE NUMBER _____

CLIENT/PATIENT INFORMATION FORM

Please provide the following information for our records: **PLEASE PRINT!**

OWNER INFORMATION

Owner's Name	Social Security Number
Street Address	

City/State	Zip Code	County

Telephone (Include Area Code)	Home	Business

Driver's License Number	Place of Employment	How Long Employed?

ANIMAL INFORMATION

Animal Species (Dog, Cat, Other)	Breed

Animal's Name	Sex	Has the animal been altered? □ YES □ NO

Color	Birth Date	THE UNDERSIGNED OWNER OR AGENT CERTIFIES THAT THE HEREIN DESCRIBED ANIMAL HAS A MAXIMUM VALUE OF APPROXIMATELY $.____

REFERRAL INFORMATION

Were you referred by a veterinarian? □ YES □ NO	If so, complete the following information.

Veterinarian's Name	Phone
Street Address	

City/State	Zip Code

You will be advised of estimated cost and anticipated procedures. Please feel free to discuss the proposed treatment and any costs with the veterinarian. A minimum deposit of 50% of the initial estimated charges will be required for hospitalization of the patient.

STATEMENT OF OWNERSHIP AND CONSENT: I am the owner of the above-described animal, or have authorization from the owner to consent to its treatment.

I hereby authorize the performance of professionally accepted diagnostic, therapeutic, anesthetic, and surgical procedures necessary for its treatment.
I accept financial responsibility for these services.
I have read the above consent and understand why these procedures may be necessary. I have also been told of the possible complications and alternatives to the anticipated procedures.

PAYMENT CHOICE: □ Cash   □ Check   □ Bank Card

SIGNATURE (Owner/Agent)	DATE

FIGURE 3-2
A client/patient information form is an important part of the medical record.

There are many different styles of medical records. Some practices use hanging pockets with pages inserted. These pockets can range in size from a half-page to a full 8½- by 11-inch page. Some practices use 5- by 7-inch file cards stapled together with additional information, such as laboratory reports, filed elsewhere. A veterinary medical record card might look like that shown in Figure 3-3.

The most popular medical record format is a file folder with a folding clip at the top to hold 8½- by 11-inch pages in place. This design is very easy to use, whether making additional entries or retrieving information. There are different ways to identify the client and the patient on the outside of the folder.

Much information, both objective and subjective, is contained in medical records. **Objective information** includes factual, measurable data, such as an animal's weight (Figure 3-4), body temperature, heart rate, respiration rate, results of laboratory analyses, interpretation of radiographs, or results of an electrocardiogram.

Subjective data within a medical record are entries that describe that animal's overall attitude. "Bright, alert, and responsive" is a phrase often included in this section of a medical record. Subjective information about a

Client Name _____

Address _____ Zip _____ Phone _____

Employment _____ Phone _____ Referred By _____

Animal's Name _____ Color _____

Color _____ Species _____ Breed _____ Age _____ Sex _____

Vaccination History   Date

				Allergies	_____
				Fecal Exams	_____
					_____
					_____
					_____
					_____

Date	Chief Complaint	Exam	Treatment/Surgery	Doctor

FIGURE 3-3   Veterinary medical records are often kept on 5- × 7-inch file cards

FIGURE 3-4 An animal's weight is an example of objective information that should be included in the medical record. *(Setting courtesy of Airpark Animal Hospital, Westminster, MD)*

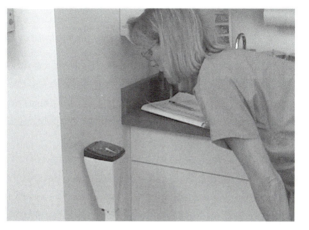

patient is not measurable with the same detachment as the objective variables. The subjective entries comprise the overall clinical impression of a patient, answering questions such as "How does the animal move?" and "How does the animal respond to its environment?"

The assessment of a patient, as well as formulation of the diagnostic and treatment plans, revolves around the careful compilation of objective and subjective data about the patient. We will look at different ways of organizing that data in the next section.

Organizing Medical Records

There are two primary ways to organize medical records that you will encounter in veterinary practices: the **source-oriented medical record** and the **problem-oriented medical record**. Regardless of the particular format, it is best to arrange information in the chart in reverse chronological order, with the most recent information on top, where it is most readily available.

The source-oriented medical record, also called the **conventional format**, can vary dramatically in size and detail. Data is entered as acquired, in chronological order. Often with this format, the records of multiple patients are kept in a single folder. Data for the different animals may be recorded on a single sheet, simply dated to indicate when each individual has been seen. This format requires much less time to complete than the problem-oriented format. However, compared to the problem-oriented system, it lacks the detailed documentation of procedures and their results, and it makes retrieving information more difficult. Maintaining adequate communication and legal protection is more challenging with source-oriented medical records. In contrast, the problem-oriented medical record format provides a complete, accurate, detailed chart on each and every patient. This format was adapted for the veterinary profession from human medicine. It offers a comprehensive review of the patient, including:

- The patient's history
- The patient's medical problems

- Procedures or treatments that have been performed and the results
- An articulated plan for the patient's future

It is more time intensive to keep this style of medical chart. However, it improves communication among staff caring for the animals, as well as among doctors in a multidoctor practice. In addition, more historical information and other data are readily available to support case planning and provide legal protection if a claim arises.

The problem-oriented medical record usually includes several standard sections. The first is the **master problem list** (Figure 3-5). Any issue requiring veterinary medical attention should be included on the problem list. Entries on the problem list might include an abnormality that the owner has noticed, an unusual laboratory finding, or a diagnosis. You can also think of the problem list as the medical history in brief or as an index to the rest of the medical chart. On the problem list, you may or may not find dates of inoculation, results of heartworm tests, drug allergies, ongoing medical conditions, and other information, depending on the standard practice of the office or hospital.

In addition to the master problem list, you may expect to find a place to record a comprehensive history, a form for recording the results of a physical examination, inserts for the results of laboratory analyses or special procedures, and progress notes (Figure 3-6). Progress notes are divided into four sections represented by the acronym **SOAP**:

- S—subjective
- O—objective
- A—assessment
- P—procedure or plan

Each problem is "SOAPed" separately. All communications with a client or any other veterinarian involved in an animal's care should be recorded in the progress notes.

The key to organizing medical records efficiently is to learn the format used in the practice where you work, become familiar with the details of the medical record-keeping system that is in place, and do all you can to help maintain the consistency of medical record keeping. This consistency ensures that each animal gets the best care by keeping all pertinent information about patients readily available for review.

Maintaining, Retaining, Purging, and Releasing Medical Records

One of the most important aspects of medical record keeping is accuracy. It is critical that each medical record always reflects up-to-the-minute information about a patient and a client. Here are some simple strategies for keeping medical records current:

- Ask the client, when she is in the clinic, if the address and telephone number recorded in the medical record are still correct.

CITY ANIMAL HOSPITAL
Master Problem List

OWNER INFORMATION

Owner Name   □ Mr.   □ Miss   □ Ms.   □ Mrs.   Patient/Pet's Name

Address   City/State/Zip

Home Phone   Business Phone

PATIENT/PET INFORMATION

Chart # _____
Patient _____   Species _____   Breed _____
Color _____   Sex   □ F □ M □ N   Birth Date _____
Vax History _____   Weight _____

IMMUNIZATION/PREVENTATIVE RECORD

DATE									
RABIES									
DA2PL									
FVR-CP									
FELV									
FECAL									

PROBLEM LIST

PROBLEM	DATE ENTERED	DATE RESOLVED
1.		
2.		
3.		
4.		
5.		
6.		
7.		
8.		
9.		
10.		
11.		
12.		
13.		

FIGURE 3-5   Issues requiring veterinary medical attention should be listed on the master problem list.

CITY ANIMAL HOSPITAL
Progress Notes

OWNER INFORMATION

Owner Name ☐ Mr. ☐ Miss ☐ Ms. ☐ Mrs. Patient/Pet's Name

Address   City/State/Zip

Home Phone   Business Phone

PATIENT/PET INFORMATION

Chart # _____
Patient _____ Species _____ Breed _____
Color _____ Sex ☐ F ☐ M ☐ N Birth Date _____
Vax History _____ Weight _____

CURRENT PROBLEMS

Date				Soap		
Month	Day	Year	Time	Format	Progress Notes	Fee
				S	Subjective Data	
				O	Objective Data	
				A	Assessment	
				P	Procedure for Diagnosis and Treatment	

FIGURE 3-6   Sample progress notes.

- Ask the client about animals not seen in a while. ("Mrs. Jones, we have not seen Fluffy since 1998. Do you still have her?")
- Record information in the medical record in a timely fashion. The day an animal is presented to the practice, what is observed and what is done, as well as any assessments and plans, should be recorded in the medical chart.
- Record telephone conversations with clients in the medical record.
- Remember to place sheets in the record with the most recent information on top.
- Allow the master problem list to serve as an index or table of contents for the rest of the record.

Periodically, it is useful for the practice to **purge**, or eliminate, the records of animals that are no longer active patients of the practice. This purging process reduces the clutter of a medical records file filled with the charts of animals that have died, whose owners have moved, or that are no longer served by your practice. When a medical record has been eliminated from the active patient files, it will be either inactivated or deleted.

Every practice has its own policy on purging, but it is generally done at least once a year. Each practice also defines which records are to be considered active. For some practices, any animal that has been seen within three years is considered an active patient. Sometimes the time frame is as short as 18 months.

If an animal has died or the owner has moved away, the file can be deleted from the active records. The hard copy of the medical chart should then be stored by the practice in case there is ever any reason to retrieve information about the animal or the case. If the animal simply has not been seen for a long time, but is not dead, the medical record should be considered inactive, but not yet deleted. Inactive files are best stored where they can easily be retrieved if the client wishes to bring the animal back to the practice for medical care.

Prior to inactivating a medical record, most practices contact a client by phone or in writing about an animal that has not been seen for a certain period of time. A contact letter might read as shown in Figure 3-7. A letter like this gives a client a chance to respond. The animal may have died or may be living in a different household. Perhaps the client is simply unaware that so much time has passed since the animal was in for veterinary care.

The standard time for storage of purged medical records, whether they are inactive or deleted, is seven years. After that time, the charts may be discarded.

A veterinary medical record is considered a confidential document. The record is owned by the veterinary practice, not by the owner of the animal. Most practices require a signed authorization form (Figure 3-8) from the owner before any information from the medical record is released to her, to another veterinarian, or to an insurance company. The attending veterinarian is usually the only person authorized to release information from the medical record.

FIGURE 3-7 A contact letter can be used to update the status of a patient that has not been seen for a while.

CITY HOSPITAL
ANY STREET
ANYTOWN, SS 00000

Dear Pet Owner:

We are updating our medical records. _____ has not been in

to see us since _____ .

If we do not hear from you by _____ , we will remove

_____ 's medical chart from our active patient records.

Thank you for your prompt attention to this matter.

Sincerely,

City Hospital

Occasionally a request will come from a public agency, such as a local animal control office, for the vaccination status of an animal that has been involved in a bite injury to a human. (This is an important exception to the general rule barring the release of confidential veterinary medical information. If you have any questions about releasing medical records in a particular situation, consult the veterinarian.) Figure 3-9 shows a sample form summarizing the vaccinations given and any other procedures performed.

When the medical record is being transferred to another veterinarian, it should be sent directly to her. The original of the medical record is never released—only copies. When the transfer of a medical record is requested, that request should be recorded in the original medical chart.

Business Records and Logbooks

By and large, the business records for a veterinary practice are kept separate from the medical records, although there is a bit of overlap in some cases. For instance, veterinary-specific software now links invoicing to the medical record and history. But most practices do their accounting using a computer process separate from their management of medical information.

Computerized business records save time and provide an enormous volume of useful information. They empower a practice to identify trends within the practice, track client purchasing choices, control inventory, and evaluate income versus expenses. Although you may be asked to learn to process invoices for individual transactions with clients, most business records are attended to by a practice manager, a business/office manager, or a practice bookkeeper.

CONSENT TO DISCLOSURE OF MEDICAL RECORD

Waiver of Confidentiality

By Authorized Patient Representative

I, _____(name of owner or agent)_____ , the _____(owner or agent of owner)_____

of _____(name of animal)_____ , a _____(breed)_____ ,

do understand that the information contained in _____(name of animal)_____ 's medical record

is confidential. However, I specifically give my consent for _____(name of veterinary practice)_____

to release the following information concerning _____(name of animal)_____

to _____(name of party to whom information is being released)_____ .

The above-listed information is to be disclosed for the specific purpose of

_____

_____

It is further understood that the information released is for professional purposes only.

This information may not be given in whole or part to any person other than that stated above.

_____

Signature of Authorized Representative

_____

Date

FIGURE 3-8   An authorization form is required to release medical records to the owner, another veterinarian, or an insurance company.

CITY ANIMAL HOSPITAL
Vaccination Record

OWNER INFORMATION

Owner Name   □ Mr.   □ Miss   □ Ms.   □ Mrs.   Patient/Pet's Name

Address   City/State/Zip

Home Phone   Business Phone

PATIENT/PET INFORMATION

Chart # _____
Patient _____   Species _____   Breed _____
Color _____   Sex □ F □ M □ N   Birth Date _____
Vax History _____   Weight _____

VACCINATIONS

PUPPY
DHP-M
PARVO
DHP/PARVO/CORONA

DOG
DHP/PARVO/CORONA
RABIES # _____
BORDETELLA
HW CHECK _____

CAT/KITTEN
FVRC-P/CHLMY
RABIES # _____
FeLV
FECAL _____

CANINE

_____ Behavior   _____ Fecals
_____ Housebreak   _____ HW
_____ Other   _____ Diet
_____ Spay   _____ Ears
_____ Castrate   _____ Nails/Anal sacs
_____ Fleas   _____ Groom
_____ Dental   _____ Microchip
_____ Lepto   _____ Bordetella

FELINE

_____ Dental   _____ FeLV
_____ Spay   _____ Declaw
_____ Castrate   _____ Chlamydia
_____ Ear Mites   _____ Fecal
_____ Fleas   _____ Diet
_____ Inside   _____ Groom
_____ Outside   _____ Microchip

TEMP. _____   C.V. _____
WT. _____   m.m. _____
GI _____   heart _____
   teeth _____   h.w. _____
   fecal _____   INTEG. _____
   anal sacs _____   ext. par. _____
RESP _____   nails _____
   upper _____   LYMPH _____
   lower _____   GEN. UR _____
EYES _____   MUS-SKEL _____
EARS _____   C.N.S. _____
   _____   ABD. PALP _____

FIGURE 3-9   Sample vaccination sheet. Note: DHP-M, M = measles, C.V.= cardiovascular, m.m. = mucus membranes, ext. par = external parasites.

Part of the business record-keeping system in any veterinary practice consists of various **logbooks**, where medicine and business overlap. One example of a logbook used in veterinary medicine is the **x-ray logbook** (Figure 3-10), which includes the following information:

- client's name
- patient's name
- date
- x-ray case number
- body part to be radiographed
- view(s) taken
- measured thickness of the body part(s) radiographed
- x-ray machine settings (kVp, MA, etc.)
- any comments (quality of the films, diagnosis, abnormalities noted, etc.)

Another example of record keeping is the **anesthesia/surgery logbook** (Figure 3-11). This volume documents the date, the patient and client, the procedure(s) performed, all drugs administered (including the exact volumes given and the routes of administration), the length of the anesthetic event, the length of the procedure, and the identities of the surgeon(s) and anesthetist.

Every veterinary practice is also required by law to have a **controlled substance logbook**. The controlled substance logbook is important from the perspective of preventing drug abuse and requiring accountability of drug use. Some medications used by veterinary practices have the potential to be abused and are therefore regulated by the Drug Enforcement Agency (DEA), a branch of the federal government. The DEA has the authority to perform random inspections, including examination of the controlled substance log.

All details of controlled substances must be recorded in the controlled substance logbook. First, the delivery of an unopened order to the central supply of the practice's pharmacy is recorded in one part of the log. Then, when a container is opened to dispense medication to a patient, that action is recorded on a different page of the log. Every tablet or capsule of every controlled substance must be accounted for in the logbook. Every milliliter of a controlled injectable medication must also be recorded.

The controlled substance logbook should include the following four items:

1. drug dispensed
2. date, client's name, and patient's name
3. amount dispensed and the amount remaining
4. names of veterinarians authorizing the use of the medication and the individual actually handling the medication

Inventory is required every two years, and the amount remaining in the logbook must agree with the amount in the controlled substance cabinet. Medical record entries should also agree exactly with the information in the

X-RAY LOGBOOK

Date	Case Number	Patient's Name	Client's Name	Body Part	View(s) Taken	Thickness of Body Part	X-ray Machine Setting

Comments:

FIGURE 3-10   A sample x-ray logbook.

ANESTHESIA/SURGERY LOGBOOK

Date	Patient	Client	Procedure(s) Performed	Drugs Administered	Dosage	Route(s) of Administration	Length of Procedure	Surgeon(s)	Anesthetist

Comments:

_____

_____

_____

_____

_____

_____

_____

_____

FIGURE 3-11   A sample anesthesia/surgery logbook.

controlled substance log. Controlled substances are to remain locked up at all times in their cabinet, which should be mounted permanently somewhere in the practice. The only staff members of the practice who should have a key to the controlled substance cabinet are the veterinarians involved in the practice.

Most veterinary facilities have a rule: "If you use it, you log it." This rule simply means that if you are the one instructed by a veterinarian to retrieve a controlled substance for dispensing purposes or for use on a hospitalized patient, you are the one who is to record that event in the controlled substance log. Therefore, you must learn and become comfortable with the process of recording the use of these medications. *Falsifying the controlled substance log or improperly using any controlled substance is a federal offense.*

The final example of a logbook used in veterinary practice is the **laboratory logbook**, which provides a synopsis of the laboratory tests that have been ordered. Usually, the results of the tests are not recorded in the laboratory log, but some practices use it that way. At the very least, the date, client and patient names, tests performed, and whether the tests are run in-house are recorded in the laboratory logbook.

FILING

Filing in a veterinary office is much like filing in any other office setting. The details vary slightly from practice to practice, but the concepts are universal. Related materials (medical records, for example) are filed in the same location (Figure 3-12). The invoices for products and materials purchased by the prac-

FIGURE 3-12 Typical medical records area. *(Photograph courtesy of Robin Downing, D.V.M.)*

tice, as well as other expenditures, are filed in a different location than the medical records—most likely in the practice's business office. The bookkeeping records from client transactions also are filed together for easy access in case they need to be reviewed. Client information and handouts are usually filed in a different location. Imprint cards for radiographs are typically found in card files near the x-ray darkroom.

Filing Equipment and Supplies

Most practices have a variety of filing equipment available, depending on the items that need to be filed. For instance, if medical records are kept in manila end-tab folders (designed to hold 8½- by 11-inch pages), there will be open shelving to accommodate the files. If the medical records are kept in hanging pocket folders, there will be cabinets with pull-out drawers.

In order to maximize the usable space within a reception area, file cabinets with lateral file drawers are often used (Figure 3-13). When the drawer of a lateral file is pulled out, it does not take up as much room as a standard file cabinet drawer. The files are then arranged from side to side rather than from front to back. If medical records are kept on index cards, you will find file cabinets with pull-out drawers of that particular size. You may also find Rolodex systems of various dimensions in use.

FIGURE 3-13 Lateral file cabinets in a veterinary reception area. *(Photograph courtesy of Robin Downing, D.V.M.)*

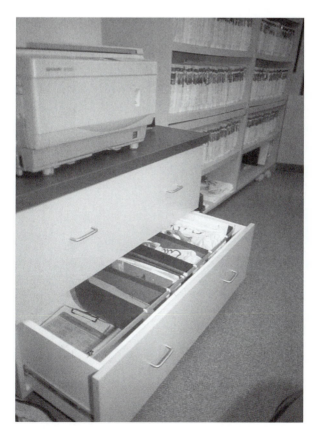

The filing supplies found in a typical veterinary practice will vary as well, depending on the filing needs of the particular practice. Most medical records are identified on the outside of the folder or pocket using letters, numbers, or a combination of the two. Therefore, you will find blank folders (usually with end tabs) or hanging pockets, rolls of number stickers, and rolls of letter stickers. In addition, you will probably find side-tab manila folders for more conventional filing. The manila folders may rest directly in a file drawer, or they may be inserted into hanging files. There will be stickers for identifying the manila folders. You will also find blank cards for imprinting radiographs, as well as blank cage cards for identifying hospitalized patients.

Filing Systems

You may encounter different filing systems, depending on the practice where you work and what kind of information retrieval systems the practice has developed. Medical records are usually filed in one of two ways: numerically or alphabetically. A **numerical filing** system can work in one of two ways: either each client is assigned a number, or each individual patient is assigned a number. Each digit, from 0 through 9, is represented by a different color. The colors on the records then vary, depending on the number that has been assigned to a particular client or patient.

As an example, a practice using numerical filing of medical records might use end-tab folders and apply the colored numbers to the end (or bottom) of the folder (Figure 3-14). Usually a single sticker with the last two digits of the

FIGURE 3-14   End-tab folders may be combined with color-coded stickers to make filing more efficient.

year is applied somewhere on the same tab. Then the medical records are filed on open shelves with the end tabs, complete with numbers, visible to the receptionist. It is easy to see at a glance if there are files out of sequence. Tabbed dividers are very helpful in clustering together records with similar number sequences.

For medical records filed using the **alphabetic filing** system, the concept is very similar to numerically sequenced medical records. Each letter of the alphabet is assigned a color, and the first two or three letters of the last name are attached to the end tab of the medical record folder. Thus, it is easy to identify if a chart is out of sequence on the file shelf.

The practice where you work may also use color coding to signal records that are to be inactivated, records for clients that are to pay cash only, records for animals that have died, records for animals that are returning for a follow-up visit, or to indicate other information. A color-coding system can be as simple or complex as the practice wishes and can be used to communicate many different types of information to others on the staff.

The practice where you work will probably have filing systems in place for more than just the medical records. For instance, your veterinarian may have client education handouts that are commonly used in the practice. There must be a single, simple way to access information like that. There will also be various forms and certificates, which are used throughout the practice, that must be kept organized and easy to find. One strategy is to have hanging files organized by topic, and then, within the hanging files, to have manila-tabbed folders organized alphabetically. This system allows related information or paperwork to be stored in the same place and to be organized and accessed easily.

Making Filing Efficient

When you are getting ready to replace files in their appropriate places, use your time as efficiently as possible. The first step is to separate the files into their appropriate groupings. For instance, gather all the medical records together, all the client education handouts together, and so forth. Next, be sure that all the pertinent and updated information is in each of the files. You should be sure, also, that all appropriate entries have been made in the medical records before they are returned to their shelves (Figure 3-15).

When you are confident that updating is complete, you are ready to begin the filing process. If you are filing medical records folders, arrange them in the appropriate order, either numerically or alphabetically, before filing them. One simple time-saving trick when retrieving a file folder from open shelving is to pull out the adjacent file a short distance. This practice marks the area from which you have retrieved the file and makes it easier to return. It is best to file medical records in a timely fashion to eliminate clutter and to keep the records available for the veterinarian's use.

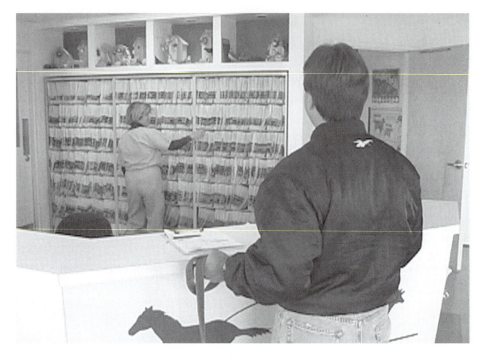

FIGURE 3-15   Make sure that medical records are properly updated before placing them back on the shelves. *(Setting courtesy of Airpark Animal Hospital, Westminster, MD)*

ADMISSIONS AND DISCHARGES

Most admissions to and discharges from the veterinary practice are conducted by the veterinarian. You may be asked, however, to assist with or to complete an admission or discharge if the veterinarian is called away. Most practices have clearly outlined procedures to follow—usually a step-by-step process that includes specific paperwork (Figure 3-16). It is important to document all procedures that are to be performed on an animal, as well as any medication the animal is to receive upon admission for a procedure.

Equally important is having written consent to proceed with diagnostic procedures or with a treatment plan. *Upon discharge, it is vital to have written discharge orders with clear follow-up instructions when the patient leaves the hospital.* This practice allows the client to review what she has been told at the time of discharging the animal and helps to prevent any confusion from inaccurate recollection of verbal instructions.

Admitting a Patient

Animal patients are **admitted** to a veterinary hospital for many reasons:

- The owner may be unable to stay for a procedure that would normally be performed on an outpatient basis, so the animal is admitted to the hospital.

FIGURE 3-16 Proper documentation is important for both admissions and discharges. *(Setting courtesy of Airpark Animal Hospital, Westminster, MD)*

- Perhaps the animal is scheduled for a wellness examination, or perhaps it is sick or injured. (Since most veterinarians prefer that the client is present during an examination, it is especially important for you to get as much pertinent information as possible from the client.)
- Perhaps the animal is being admitted for a period of supervised care while the client is out of town.
- The animal may be presented for a surgical or dental procedure.
- Perhaps the animal is boarding at the veterinary hospital and having a medical or surgical procedure done during its stay.
- Certain laboratory analyses may take more time than those performed on an outpatient basis while the client waits.

Regardless of the reason, it is important to know precisely why the animal is to be admitted to the practice.

If the doctor is unavailable to admit the patient, then it is critical that you get all the relevant information you can from the client at the time of the admission. You must be sure to take a thorough and accurate history. The systems history, environmental history, and systems review are all important. Record any specific concerns or questions the client has. Also, be sure to ask where and when the client may be reached. Record telephone or pager numbers in case the client must be reached to answer questions, to give consent for additional procedures, or simply to be updated about the animal's condition.

An animal should never be left at a veterinary practice without a written, signed **consent form** in the medical record. By reading and signing the form, the client gives his or her informed consent for the procedures to be done and acknowledges the potential risks and benefits. You may be called upon to prepare consent forms. Most hospitals use either preprinted fill-in-the-blank forms, or they have prepared consent forms, loaded into the computer system, that may be personalized with specific client and patient information (Figure 3-17). Most forms tend to have fairly standard consent language. You will need to become very familiar with the forms used by the veterinary practice where you work, because you may be asked to answer clients' questions about giving their consent.

You should be able to walk the client through, step by step, each section of a consent form and each procedure that is described. For example, on a surgical consent form, there will probably be various preoperative safety precautions recommended by the veterinarian, including the benefits and risks of anesthesia. The form may include precautions to be taken during the operation, such as intravenous fluid therapy. There may be an entry for postoperative pain management such as tranquilizers and analgesics. You need to know what your veterinarian recommends and why. When the consent form is signed, the client gets a copy for his or her own records, and the original goes into the medical record.

For a nonstandard admission (for example, a patient who needs an internal medicine workup), the veterinarian will list on the consent form all procedures she anticipates doing with the animal. Sometimes, however, you will find that the veterinarian uses general language like "appropriate laboratory tests." Once again, the client gets a copy of the signed consent form, and the original is inserted into the medical record.

Discharging a Patient

Most often, a veterinarian will **discharge** a patient after a period of hospitalization in order to review with the client test results, surgical findings, postoperative care, or recommended follow-up instructions (Figure 3-18). Occasionally, however, you may be asked to assist with the discharge of a patient, or you may be asked to conduct the discharging procedure on your own. You should be familiar with the process.

Your practice may use standard written discharge instructions to follow what are considered routine procedures. These instructions may be stored in the computer to be personalized for each patient, or they may be preprinted forms. Most discharge orders have blanks to be filled in and most include comments about the following:

- prescribed medications, including the drug name, the dosage, and how often it is to be given
- dietary modifications
- the animal's recommended activity level
- bathing or being in water, playing with other pets, or the level of necessary confinement
- an incision if one is present

(continued on page 63)

**CITY HOSPITAL
ANY STREET
ANYTOWN, SS 00000**

CONSENT FORM

Owner's Name: _____   Animal's Name: _____

Address: _____   Species: _____

_____   Breed: _____

Case Number: _____   Sex: _____

I am the owner or agent for the owner of the above-described animal and have the authority to execute this consent.

I hereby consent to the hospitalization of the above-described animal and authorize the veterinarian and staff to administer any tests, medications, anesthesia, or surgical procedures that the veterinarian deems necessary for the health, safety, and well-being of the animal.

I specifically request the following procedure(s) or operation(s):

_____

I understand that during the course of the above-mentioned procedure(s) or operation(s), unforeseen conditions may be discovered that necessitate an extension of the above-mentioned procedure(s) or operation(s) or additional procedure(s) or operation(s). I hereby consent to and authorize the performance of such procedure(s) or operation(s) as are necessary according to the veterinarian's professional judgment.

I also authorize the use of appropriate anesthetics and other medications. I understand that the veterinary support personnel will be employed as necessary according to the veterinarian's professional judgment.

I have been advised as to the nature of the procedure(s) or operation(s) to be performed and the risks involved. I realize that results cannot be guaranteed.

I understand that all fees for professional services are due at the time of discharge.

I have read and understand this authorization and consent.

Additional Comments/Information:

_____

_____

_____

Date _____   Signature of Owner or Agent _____

                                Signature of Witness _____

FIGURE 3-17   No animal should be left at the veterinary practice without a signed consent form.

CITY HOSPITAL
ANY STREET
ANYTOWN, SS 00000

FOLLOW-UP INSTRUCTIONS FOR HOME CARE AFTER SURGERY

Due to the effects of general anesthesia, your pet may appear more tired than normal and possibly a little uncoordinated. This is to be expected, and the grogginess should disappear in a day or two, at the latest.

- To prevent vomiting due to excitement upon arriving home, do not give your pet food or water for an hour after returning home. Feed the regular diet lightly today, then return to normal feeding tomorrow.

- If your pet becomes listless or refuses to eat the next day, or if vomiting or diarrhea occurs, call the veterinary hospital immediately.

- Discourage your pet from licking or chewing the stitches or incision. Call us if licking or chewing persists.

- Check the incision site twice a day for any redness, swelling, pain, or drainage. Report any of these immediately.

- Limit exercise for a week following surgery. (No running or jumping)

- Keep any bandage or cast clean and dry. Contact the veterinary hospital if the bandage is excessively wet or dirty.

FUTURE TREATMENT

☐ Please make an appointment for a recheck in _____ days/weeks.

☐ Please make an appointment for suture removal in _____ days.

☐ Please make an appointment for bandage change or removal in _____ days.

☐ Remove bandage at home in _____ days.

ADDITIONAL INSTRUCTIONS:

_____

_____

_____

_____

_____

_____

FIGURE 3-18 It is important to provide follow-up instructions for animals being discharged from the veterinary practice.

- specific instructions about what to look for and when to call if a problem arises
- scheduled follow-up visits

When the client has reviewed the discharge orders and has had his or her questions answered, a copy of the signed discharge form goes with the client and a copy remains in the medical record. At the time of discharge, the client often receives a copy of any laboratory analyses that have been performed.

FORMS AND CERTIFICATES

You will find many different forms and certificates in common use at a veterinary practice. Become familiar with all of these in case you are asked to fill one out or to answer a client's questions.

We discussed, in the previous section, consent forms and discharge instructions. Other forms may be created for the client at the time a patient is discharged from the hospital. A **neuter or spay certificate** allows the client to prove that the animal is no longer sexually intact. This proof may allow the client to get a discount on the pet's license and may be required for the client to fulfill his or her obligation to a shelter from which the animal has been adopted.

Vaccination certificates are usually prepared any time an animal has received an inoculation at a practice (Figure 3-19). These certificates may be handwritten on government-approved standard forms, or they may be prepared on the practice's computer system—linked to invoices so that they are automatically printed any time the animal receives a vaccination. Computer linking can also generate future mailing notices for inoculations, as well as help identify a lost animal.

Health certificates, also commonly used in veterinary practice, are more complex forms that are required for interstate and international transport of an animal. In most cases, the veterinarian is the only person authorized to sign a health certificate for travel.

Most certificates used in veterinary practice require a veterinarian's signature. If the form is not required to be filled out by only a veterinarian, then you may be asked to prepare the certificate ahead of time or while the patient is in the hospital. In these cases, you will present the form to the veterinarian for his or her signature when the information contained on it is as complete as possible. You may be asked to place the certificate on the doctor's desk or in an internal mailbox or message box, or you may simply be asked to have the form ready for signature at the front desk when the veterinarian escorts the client to the reception area.

Every practice has a slightly different routine to be followed for ensuring that record keeping, certificates, and other appropriate paperwork are handled in a timely, efficient, and accurate manner. It is up to you to learn the details of how things are done at the particular practice where you work.

**CITY HOSPITAL
ANY STREET
ANYTOWN, SS 00000**

RABIES CERTIFICATE

Patient Information

Client Name: _____   Patient Name: _____

Address: _____   Species: _____

_____   Breed: _____

Telephone (H): _____   Sex: _____

(W): _____   License No.: _____

Vaccination Information

Vaccination Date: _____   Expiration Date: _____

Vaccine Serial No.: _____   Producer Killed: _____

Comments:

_____

_____

_____

_____

Signature of Veterinarian

FIGURE 3-19 Sample rabies vaccination certificate.

SCREENING AND PROCESSING MAIL

In your position as a veterinary assistant, you may be asked to process the practice's mail, both incoming and outgoing (Figure 3-20). You need to know the basic elements and be prepared to ask for the specifics as they apply to your particular practice.

Incoming Mail

You will need to know the recommended process for screening the incoming mail in your practice. Be prepared to be able to answer the following questions:

- Is the mail delivered to the practice facility by a mail carrier, or must it be picked up at the local post office?
- Is collecting the incoming mail the specific responsibility of one person in the practice, or is it a shared job?
- What time of day is the mail delivered, or by what time can you count on the mail to be ready in the post office box?
- What time can you expect alternative deliveries like UPS or Federal Express?
- Who may sign for deliveries of mail or packages that require a signature?

For most mail, opening all envelopes with a letter opener is the best first step. It is a courtesy to have the envelopes cut open before they end up in the individual mail slots or boxes within the practice. Simply open the envelopes and set them aside to be sorted. Do not sift through the contents. Leave unopened any envelopes or packages marked "personal" or "confidential."

Next, sort the mail by **addressee** (recipient), and then place each piece into the appropriate receptacle. Does your practice have trays marked with each staff member's name for depositing phone messages and mail, or is it the receptionist's job to see that mail and messages get into the right hands? Who receives envelopes addressed to the practice (the practice owner, the clinic manager, or someone else)?

FIGURE 3-20 Many veterinary practices send holiday cards to their clients. *(Photograph courtesy of Robin Downing, D.V.M.)*

You need to learn the process for sorting through catalogs and other "junk" mail. Who in the practice decides what is "junk" mail? Is there a central location for catalogs? How are the catalogs separated? Whose responsibility is it to go through the catalogs periodically to purge duplicates and keep only the most current ones from each company? Do catalogs simply go into the internal mailbox of the person to whom they are addressed? Are discarded catalogs and other advertisements recycled? Where are they gathered so that they may be recycled?

You must also learn the process for accepting deliveries of medical or office supplies to the practice. Where do the boxes go when they arrive at the practice? Who is responsible for unloading shipments of medical supplies? Is there any way you can assist that person, either by reviewing the packing slip to be sure that everything has arrived or by helping to put things away and rotating the stock? How are packages to be handled that read "Refrigerate Upon Arrival"?

Every practice has its own ways of dealing with incoming mail and packages. It is up to you to learn the established routine and to fit into that routine wherever you can be most helpful.

Outgoing Mail

When sorting through outgoing mail, the first step is to classify it for postage and handling. First-class letters receive different postage than reminder postcards directed to clients. Packages are handled differently as well. You will need to learn the procedure for handling outgoing mail at your practice. You will need answers to the following questions:

- How is the outgoing mail postage handled?
- Is there a postage scale in the practice for weighing mail and determining the appropriate postage?
- Is there a postage meter in the practice for applying postage without using stamps and without having to take the mail to the local post office for postage to be applied?
- Is the outgoing mail picked up at the practice by a postal carrier or does it need to be taken to the local post office?
- Which person in the practice contacts UPS, Federal Express, or other alternative carriers for a pickup at the practice?
- Will you be authorized to prepare outgoing packages for pickup?

Handling Mail Problems

Handling different types of problems with letters, packages, and carrier services depends on the problem. You will be required to become familiar with the procedures followed at the practice where you work.

For instance, if a letter or a reminder postcard to a client is returned because the address is no longer correct, you need to know how to track down the new address. Sometimes the forwarding order has expired and the postal service is no longer delivering mail to the new address after it has been sent to the previous address. Very often, however, there is a sticker on the front of the card or envelope that lists the new address. It is easy to update the medical record and client file at that time. Sometimes a client receives mail at a post office box rather than at a physical address. This problem is relatively easy to avoid by carefully filling out the client information form or computer screen when the client first comes into the practice. Most new clients distinguish between the physical address and the mailing address.

If clients ask the practice to send a product or medication to them because it is inconvenient to stop in to pick it up, you need to know how to handle the details. Whose responsibility is it to decide what can be sent through the mail and what cannot? How are medicines packaged? How are the clients charged?

Most problems with postal or package services can be resolved easily by using common sense and knowing who in the practice is most likely to have the answers to your questions. You need to know who has the authority to make decisions about incoming and outgoing mail and packages. You will then be in a position to know how you can be most helpful in processing the large volumes of mail that pass through a typical veterinary practice.

Like many other aspects of working in a veterinary practice, handling mail involves some training and some common sense.

SUMMARY

Medical records are the single most vital tool a veterinarian has, providing her with the patient's medical history, allergies, and growth history. Proper documentation of information relating to clients, their animals, tests and procedures, and medications is essential from the standpoints of health, efficiency, and legality. Veterinarians rely on their administrative staff to ensure that medical records and communications are properly labeled, filed, and organized in an efficient, easy-to-understand manner.

The administrative end of the veterinary practice is, in many ways, just as important as the medical end. As a veterinary assistant, you will likely be responsible, at least in part, for maintaining business records and logbooks, forms and certificates, admitting and discharging patients, and even handling incoming and outgoing mail. While these duties may seem mundane at times, even the slightest error can have far-reaching consequences, so it is essential that you are knowledgeable and skilled in performing your duties.

CASE STUDY

Mrs. Morrison brought her cat, Muffin, in to see Dr. Rockwell for a strange lump she found on its stomach. Sandy, the veterinary assistant, greeted the very upset woman in the reception area, quickly pulled Muffin's file and showed them into the examination room to meet with the veterinarian.

Sandy held and petted Muffin while Dr. Rockwell reviewed the file and asked Mrs. Morrison, who seemed certain it was cancer, some questions about the cat's recent health. The vet noticed in the file that Muffin had been brought in several years ago by Mr. Morrison for a similar problem, which turned out to be a fluid buildup from a minor infection.

On examining the cat, Dr. Rockwell quickly deduced that the Muffin did indeed have another infection. He explained the situation to Mrs. Morrison, who then remembered being out of town at the time, when her husband had brought the cat in. The vet drained the fluid and prescribed the same antibiotic used for the last infection, assuring Mrs. Morrison that Muffin would be just fine in a few days.

- How did the use of medical records play into this scenario?
- What might have happened if the medical record had not been available?
- How did proper filing aid Sandy in this situation?

REVIEW

Indicate whether statements 1–7 are true or false.

1. Medical records provide for rapid retrieval of client information as well as patient information in case of an emergency.

2. Medical records allow a veterinarian to track certain medical trends within the patient population.

3. All information in a medical record should be written in pencil in case important changes need to be made.

4. After the client signs the consent form, the original should be given to the client, and a copy should stay in the medical record.

5. Upon admission of his or her animal, a client can give verbal consent for the staff to proceed with diagnostic procedures and treatments.

6. A surgical consent form should include an explanation of the benefits and risks of anesthesia.

7. When sorting mail, you should open all envelopes with a letter opener, including personal or confidential mail.

8. List 10 components of a medical record.

9. Who is authorized to release information from a medical record?

10. How long should inactive or deleted medical records be stored?

11. List three precautions that you must take when recording information in the controlled substance logbook.

12. What are the two systems commonly in use for filing medical records?

13. There are three different types of medical records mentioned in the text. What are they?

14. List three ways to make filing more efficient.

15. List five considerations for a patient that a discharge form might address.

16. Name three common, but important, types of certificates that require the signature of a veterinarian and describe their purpose.

17. List three questions you should ask to learn the procedure for screening incoming mail.

18. List two alternatives to the U.S. Postal Service for picking up outgoing mail.

ON-LINE RESOURCES

Veterinary Glossary (from About.com)

The Veterinary Glossary in the Agriculture section of About.com offers a very thorough list of veterinary terms, definitions, and abbreviations. Reviewing this list can help students familiarize themselves with the language they are apt to hear and use in the veterinary office.

<http://agriculture.about.com>
Search Term: Veterinary Glossary

Quia

This site offers a number of games designed to test a person's medical or anatomical knowledge, including one featuring medical record terms.

<http://www.quia.com>
Search Terms: Medical; Anatomical

KidsHealth

This site features "Knowing Your Child's Medical History," an article about how to keep a personal medical record of a child's allergies, medications, illnesses, and so on. Although designed for humans, the ideas discussed in this article could also apply to pets.

<http://kidshealth.org>
Search Terms: First Aid; Medical History; Parent

UNIVERSITY OF NEBRASKA - LINCOLN
Institute of Agriculture & Natural Resources
Department of Agricultural Leadership
Education and Communication
P.O. Box 830709
Lincoln, NE 68583-0709

CHAPTER 4

Computers

OBJECTIVES

When you complete this chapter, you should be able to:

- identify the basic components of a computer and explain the function of each
- explain what software is and give examples of general types of computer programs used in veterinary offices
- describe different types of computer networks and how they are used
- describe common features of veterinary practice Web sites and how they can affect the administrative function of the practice

KEY TERMS

central processing unit	zip drive	workstation
floppy disk	digital camera station	server
CD-ROM	hardware	extranet
CD-R	toner	firewall
monitor	software	Internet
mouse	operating system	Internet service provider
keyboard	word processing	database
printer	network	broadband
modem	intranet	Web site
scanner		

INTRODUCTION

Veterinary offices, like virtually all other medical practices and hospitals today, rely on computers for scheduling, organizing data, tracking patient information, accounting and billing, and so on. Although you probably have some experience working with computers, you have probably not had exposure to veterinary programs, and may not have used a computer station on a network. If you have never used computers before, they may seem a little intimidating at first. Rest assured, though, that computers and computer programs are not designed to be difficult. Once you develop a rudimentary understanding of computers, you will find it much easier to operate them. Remember, *the best way to learn how to use a computer is to practice!*

There is such a wide variety of computers and computer programs available that we cannot really specify here how to use every single one of them. On the job, you will learn to use your employer's particular software programs. But you will be expected to know some general facts about computers.

COMPUTER COMPONENTS

We start with the basic computer setup itself. In brief, the computer system you are likely to encounter in a veterinary practice or animal hospital will consist of a **central processing unit** (CPU) with a **floppy disk** drive and **CD-ROM** or **CD-R** drive, a **monitor, mouse, keyboard**, and a **printer** (or possibly multiple printers). It might also include a **modem** (if your office has Internet access), a **scanner**, external storage devices such as a **zip drive**, or a **digital camera station**. The physical components of a computer system are known collectively as **hardware**.

The printer is an especially important piece of equipment, as you will be using it to print various reports, client information sheets, and other items. You will need to know how to load the paper, change the **toner** cartridge and drum, or printer ribbon, and other basic maintenance (Figure 4-1). You should speak with your supervisor about how to perform these functions.

You will generally not need to know a great deal about the hardware beyond turning everything on and off correctly. You will need the most training for the **software**, the programs used with the computer. Software may include the **operating system**, **word processing** programs, Internet and e-mail applications, and any special veterinary programs.

VETERINARY SOFTWARE

Computer software can simplify most of your tasks in a veterinary office or animal hospital. There are a number of veterinary computer programs on the market today, used for a variety of purposes, including:

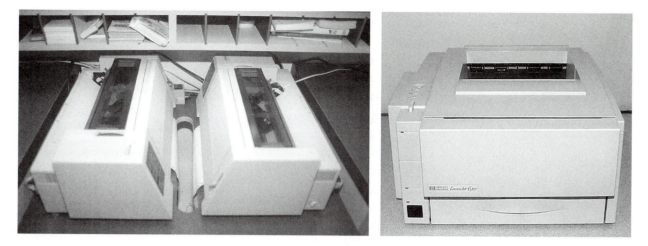

FIGURE 4-1   Your practice's computer may have one or more printers attached that you will need to become familiar with. *(Photograph on left courtesy of Robin Downing, D.V.M.)*

- aiding in medical diagnosis
- organizing medical records
- scheduling
- inventory management
- marketing
- generating performance evaluation reports
- time card processing
- posting account ledgers
- tracking accounts receivable
- separating accounts payable records
- merging new files into an alphabetical or numerical filing system
- creating invoices
- reconciling bank statements
- keeping track of inventory
- formatting, storing, and printing form letters, mailing labels, payroll summaries, paychecks, charts, graphs, bank statements, addressed envelopes, telephone directories, and other data

The ease of performing these tasks on a computer explains why most veterinary assistants feel fortunate to work in a computerized office.

Commonly Used Programs

There is no single industry standard program that you will find on every computer, but here are some common types of programs and examples of each:

- operating system: Mac OS, Microsoft Windows
- general word processing: Microsoft Word, WordPerfect

- general database management: Corel Paradox, Microsoft Access
- general spreadsheet: Corel Quattro, Microsoft Excel
- general communication (including e-mail): Corel CENTRAL, Microsoft Outlook
- Web browser: Internet Explorer, Netscape
- specialized veterinary practice/hospital management: Alis-Vet, Autovet, AVImark, BWC Animal Hospital Management System, Complete Clinic, CompuVet, DVMax, e-Friends, ImproMed, IntraVet, PAWS Veterinary Practice Management, PCVet, V/Boss, V-Tech Practice Management System, Vet Windows, Vet's Pet, Vetpak, Vetech Advantage, Vetstar Hospital Management System, Vetware
- specialized veterinary diagnosis/laboratory: Vetstar Animal Disease Diagnostic System, VisuaLab

Your office will probably use a combination of several of the programs from this list, and possibly others. Because every software program is different, it is important that you undergo specific training at the particular practice where you work. Some practices do their own in-house training of employees on the computer system, while others provide on-site instruction by outside experts in their particular software products.

COMPUTER NETWORKS

Unless you work in a very small practice, there will likely be more than one computer in the office. Veterinary hospitals and practices with multiple computers often employ a computer **network** system that links the computers together. There are three basic types of computer networks:

1. **Intranet.** An intranet is a computer network that links only the computers in a single office or building. Only authorized computer users in the building can access the intranet, using an assigned user name and password. Each computer on the network is called a **workstation**. Each workstation is wired into a central hub, which allows the workstations to "talk" to one another. When there are more than a few computers in the office, there may be a central **server**, a computer dedicated to managing network traffic and resources. The server might be used to store and manage common files, process print jobs for shared printers, and maintain a shared Internet connection.
2. **Extranet.** An extranet serves essentially the same function as an intranet, but allows limited access to people off-site, who dial in through a modem. A user name and password is also required to gain access to the extranet, which would exist behind a **firewall** (software or hardware designed to limit access to a server) on the server. Extranets are most often used by large companies with employees who

frequently work off-site or by organizations with multiple locations that need to share certain types of information.

3. ***Internet.*** The Internet is an enormous worldwide series of computer networks that allows individual computer users to access innumerable personal or business servers. Although you generally need a user name and password to gain access to the Internet through your **Internet service provider** (ISP), you generally do not need them to gain access to most features the Internet offers. However, many sites do have subscription-based services or other areas on their Web sites that limit access to certain people. For these, you would need a separate user name and password that would be assigned by that organization.

Because a veterinary practice or hospital is generally a single-location entity, it most likely uses an intranet. It probably also has Internet access.

Why would a veterinary hospital or practice use a computer network? Common uses include:

- *File sharing.* The most common use of a computer network is file sharing. Any files that several or more people need access to can be stored in a single location and pulled up as needed. This feature is especially useful when a change is made. For example, if you update a document and save it, everyone from that point on will have the most up-to-date version, whereas in the past, each individual person would have to make the same update.

- *Printer sharing.* Most offices do not have separate printers for every computer, so a network can be used to share a printer among various users. It can also be used to enable different types of printers. For example, you might print a document on a laser printer in black and white and a graph or color-coded chart on a color inkjet.

- *Sharing database access.* A **database** is a special type of file that contains uniform pieces of information organized into an accessible system. For example, your practice might have a client database with individual fields for the client's name, the patient's name, an address, phone number, and other information. Because databases are often updated frequently, shared access allows each user to have the most up-to-date information.

- *File transfers.* In the past, when you had a file on your computer that you wanted to give to someone else, you had to put it on a disk and transfer it physically. Many files are too large for floppy disks, and you might not have access to a zip disk drive or a CD-writer. Using a computer network, you can directly copy a file from your computer to another person's workstation much faster and much easier than any type of physical disk transfer.

- *Communication.* An intranet can also be used to facilitate communication between veterinary employees, especially in larger hospitals or offices and when each person has his or her own workstation or computer account. Telephone messages, memos, and quick notes (i.e., "Dr. Peterson, the lab results are back and ready for your review.") can be sent via the network in the form of e-mail or virtual memo notes. Such notes are more reliable than pieces of paper because they will not get lost or accidentally thrown away, and only the intended recipient will see them.

- *Sharing Internet access.* If your practice is large enough to have a computer network, it most likely also has a **broadband** Internet connection, which is a connection that is constantly on and, with the help of the network, can be shared by any or all computer users at the same time. A broadband connection is also much faster than a typical dial-up connection using the telephone line.

THE VETERINARY PRACTICE WEB SITE

It is becoming more and more common for veterinary practices and hospitals to have an Internet presence in the form of a **Web site**. Like most other business Web sites, it contains basic information such as the name of the practice, the address and phone number, office hours, a brief description of the practice, and the names and credentials of the veterinarians. Other common components of a veterinary Web site might include:

- driving directions and a map
- an e-mail link so that visitors may ask questions
- links to related sites, such as general animal-interest sites, information about particular breeds, veterinary news sites, or others
- an on-line library of articles, definitions, common ailments
- on-line appointment scheduling, which would involve clients entering their information into a form and requesting a certain day and time to come in, and then having a representative contact them to verify or change the appointment
- prescription refill information and on-line request forms
- listing of specialty services (such as radiology, laser surgery, veterinary dentistry) and explanations of what is involved
- general pet health and care information
- employment opportunities

If your practice has a Web site, you might be assigned the responsibility of checking and answering client e-mails and contacting clients about on-line appointment and prescription refill requests. If so, you should be sure to incorporate these duties into your daily routine, checking the e-mail each morning when you come in, and at designated points throughout the day. You should return e-mails and other on-line communications in as timely a manner as you would respond to a telephone call or letter. Responding quickly and effectively to on-line requests reflects positively on the practice and the veterinarians, and shows that you are caring and professional.

SUMMARY

Computers have become an important mainstay in most veterinary practices, so it is important that you learn how to use them. The physical components of a computer are called *hardware*, and the programs used on the computer are called *software*. There are a number of different types of computer programs that are used for different purposes, including word processing, database management, and e-mail and web browsing, as well as specialized veterinary software used for diagnosis, medical record organization, scheduling, and other functions.

Interconnected computers are called a *network*. Most practices that use computer networks use an intranet, which allows users to share information and files, communicate on-line, and share printers and Internet access. Your practice might also have a Web site, which may allow clients to request prescription refills and schedule appointments on-line. If so, you might be responsible for keeping track of such on-line requests and communicating with clients in a timely and professional manner.

CASE STUDY

Marta is a veterinary assistant at the Sunnydale General Animal Hospital. One of her daily responsibilities is to check the e-mail each morning to see if any clients have sent in questions or prescription refill or appointment requests. She jots down the names of two clients requesting prescription refills and sets the note aside to pass along to Elvin, who is in charge of pharmaceuticals. She also sees that Mr. Atkinson sent in a request at 10:30 the previous morning, asking to have an appointment at 2:00 the next afternoon for his dog, Teddy. "Too bad," she thought, "If he had only e-mailed me earlier, I would have been able to fit him in. Unfortunately, Mrs. Sordello called yesterday afternoon asking to come in at the same time, so I gave her the appointment." Marta e-mailed Mr. Atkinson back with a list of available times, and then logged off.

- How did Marta use the computer to aid in her work?
- How could Marta have prevented the scheduling conflict?
- How else could Marta have used the computer to make her job go easier and faster?
- How should Marta use the computer again today? Why?

R E V I E W

Indicate whether the following statements are true or false.

1. The best way to learn how to use a computer is to read.

2. Your computer's monitor is considered a piece of software.

3. Your computer's keyboard is a type of hardware.

4. A laser printer uses toner instead of a ribbon.

5. Microsoft Windows is a popular word processing program.

6. If your veterinary practice has a computer network, it will most likely be an intranet.

7. Veterinary computer programs can only be used to diagnose medical conditions.

8. Broadband connections are faster than dial-up connections.

9. A computer network can be used to share printers and files.

10. A Web site can be used both to provide information to clients and to facilitate communication.

O N - L I N E   R E S O U R C E S

Webopedia

This site is an on-line dictionary of computer and Internet terms. Visitors may search by keyword or category, or review the latest additions, the top fifteen terms, or the term of the day.

<http://www.webopedia.com>

MyVetOnline Directory of Veterinary Websites

Use this site to locate veterinary practice Web sites all over the country.

<http://www.myvetonline.com>

Microsoft

Microsoft is the manufacturer of the Windows operating systems and the Microsoft Office suite of applications. This site features information regarding the latest releases, free updates, and driver downloads.

<http://www.microsoft.com>

Apple

Apple is the manufacturer of the Macintosh series of computers and operating systems. This site features information regarding the latest releases, free updates, and driver downloads.

<http://www.apple.com>

Corel

Corel is the manufacturer of the WordPerfect suite of office applications. This site features information regarding the latest releases, free updates, and driver downloads.

<http://www.corel.com>

PART II

The Veterinary Assistant

CHAPTER 5

Interpersonal Communication

OBJECTIVES

When you complete this chapter, you should be able to:

- describe the communication process and list the basic elements of communication
- list and explain five personality traits that are essential for interpersonal relations in the veterinary office
- distinguish between appropriate and inappropriate professional interactions with clients

- recognize prejudice, insensitivity, or discrimination in interpersonal relations
- list and describe at least five barriers to effective communication
- list three ways you can improve your speech

KEY TERMS

communicated	patience	verbatim
interpersonal communication skills	kindness	prejudice
communication process	tact	stereotype
message	courtesy	volume
feedback	empathy	pitch
reference points	listening	tone
relate	body language	monotone
	paraphrase	enunciation

INTRODUCTION

Think back to when you woke up this morning. What did you do? Most likely, you read a newspaper, listened to the radio, or watched television. Within an hour or so, you probably spoke to a family member. Later, you may have talked on the telephone, checked your e-mail, or gone on line. At some point in your morning, you **communicated**—you exchanged or received information—with someone.

As people, we communicate with each other all the time. However, not all of our communication is effective. For instance, did you ever try to have a conversation with someone who for some reason or another was not listening to you? Or have you ever tried to read a letter that was too difficult or written too poorly for you to understand? It is frustrating when the lines of communication are shut down.

As a veterinary assistant, you will rely heavily on your **interpersonal communication** skills, the way you communicate with other people. In your new job, you will need to communicate successfully with office staff and clients. You may handle incoming and outgoing telephone calls, schedule appointments, and greet clients. In order to be successful at your new job, you must have good interpersonal communication skills.

In this chapter, you will learn how to communicate verbally with clients, both in the office and over the telephone. You will also learn how to interpret nonverbal communication—body language, eye contact, appearance—so you can make sure the nonverbal messages you are sending to clients are positive (Figure 5-1). However, before you learn about interpersonal communication, let us make sure you understand how communication works.

FIGURE 5-1   This slouching, veterinary office manager is sending out a negative message.

THE COMMUNICATION PROCESS

In general, the **communication process** has four essential elements:

1. message
2. sender
3. channel
4. receiver

A **message** is an idea that one person (the sender) wants to get across to another person (the receiver). The person gets this message across through a particular channel of communication (writing, speaking, or even a meaningful glance). This entire system is known as the communication process (Figure 5-2). The receiver can send a return message, which is called **feedback**. The whole process is then reversed—the receiver becomes the sender, and the sender becomes the receiver.

Sounds simple, does it not? Well, unfortunately, lots of things can interfere with a person's message. To begin, all senders and receivers have their own **reference points** that determine how they express and understand messages. Education, experience, social and cultural barriers, and religious beliefs are some of the most significant reference points in our lives. So, even though you think your message is clear, a client may interpret it incorrectly just because of his or her reference points.

Other factors can interfere with a person's message, too. Outside noises or actions may distract the listener. In writing, the author's style may be confusing, he may have poor handwriting, or the reader may have poor reading comprehension.

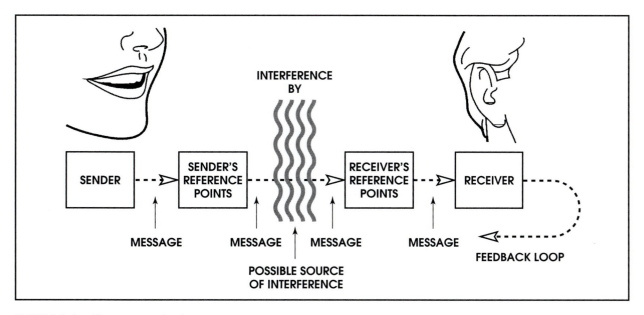

FIGURE 5-2   The communication process.

Good communicators adapt their speech to the listener's needs, expectations, and ability to comprehend. As a veterinary assistant, you must be able to "read" other people quickly, so you can send them messages they understand and appreciate. For example, some clients will have little or no knowledge of veterinary medical terminology, so you should take care to define any veterinary medical terms you use when you speak with them. Other clients will pride themselves on understanding as much as they can about their pets' health; these clients might be offended if you talk down to them.

TRAITS FOR POSITIVE INTERPERSONAL RELATIONS

Think of someone you enjoy talking to. Chances are, you have something in common with him. The two of you **relate** to each other in some way. When you become a veterinary assistant, you should try to relate to your clients. When you relate to someone, you make a connection with him. Now, do not panic. Relating to people is not as hard as it sounds. You will relate to some of your clients without even trying. With others, however, you will have to work at it a bit. Developing the following personality traits will help you relate to all of your clients:

- patience
- tact
- kindness
- courtesy
- empathy

To have **patience** means to bear trials calmly without complaint. As a veterinary assistant, you have to be patient with animals—and their owners. Not all of your clients behave as you would like them to. Some clients continually show up late for appointments. Others bring their pets into the office unrestrained, even though you may have asked them repeatedly to use a leash. Once in a while, clients will even blame you for things beyond your control, such as the length of time they have to wait to see the veterinarian. You need great patience to deal with people like this, and it is important for you to resist the temptation to react.

Impatient: I'm sorry, Mrs. Smith, but it is not my fault that Dr. Curtin had to perform an emergency surgery while you were in the waiting room.

Patient: I'm sorry that the unexpected delay has inconvenienced you, Mrs. Smith. Dr. Curtin will probably be in surgery another 10 or 15 minutes. Would you like me to reschedule you for another appointment this week?

Also, you need to resist the temptation to get clients to pay their bills faster, or to get their animals in and out of carriers more quickly.

Kindness, for the veterinary assistant, means being helpful, compassionate, and friendly. It means treating others as you would want to be treated if the situation were reversed.

> *Insensitive:* Hi, have a seat.
>
> *Kind:* Good morning, Mr. Reardon, how are you? Muffin looks very good today. Please make yourself comfortable in the reception area, and we will call for Muffin as soon as we can.

Tact means doing and saying the right things at the right time. If you are tactful, you maintain good relations with others and avoid offense. To avoid potential communication problems, do some common sense planning. A nonverbal cue such as a sign can eliminate misunderstandings leading to verbal conflicts. Directions such as "Please announce your arrival to receptionist" or simple courtesy signs such as the one pictured in Figure 5-3 can nonverbally communicate your requests. As a veterinary assistant, you need to be perceptive; you need to know your own feelings and those of others. Oftentimes, it is not *what* is said but *how* it is said that causes offense. For instance, at times you may have to remind clients that they are behind on payments or that payment is required on the day services are rendered.

> *Tactless:* Mr. Collard, you still owe $30 from Smokey's last visit, and we expect you to pay it today.
>
> *Tactful:* Mr. Collard, I just want to let you know that today's bill will include the $30 balance remaining from Smokey's last visit.

You also need to be tactful when you tell clients that their animals or their children are disrupting the reception area.

FIGURE 5-3 Sometimes you can convey messages tactfully and indirectly, without having to create a verbal issue.

Thank You
For
Not Smoking

Tactless: Mrs. Garrett, can you stop Rex from barking? He's so loud that I cannot hear the person on the other end of the telephone.

Tactful: Excuse me, Mrs. Garrett—I am sorry to interrupt. I have trouble hearing over the phone when there is background noise. Could you please take Rex into the next room for a few minutes?

Tactless: Kids, stop running around and making all that noise.

Tactful: Children, here are some magazines and books for you to read quietly.

Courtesy means putting the needs of other people before your own. It means cooperating, sharing, and giving. You should treat all clients on a polite, professional, and impartial basis. *Please, thank you, you're welcome, excuse me,* and *may I help you?* should be standard phrases in your vocabulary. And, you must be careful not to play favorites; do not do special favors for one client that you would not do for another.

Empathy means being able to feel and understand what the other person is feeling. When you empathize with your clients, you show that you understand what they are feeling.

CLIENT: I've never felt so exhausted in my life. Housebreaking a puppy is hard work.

VETERINARY ASSISTANT: Yes, training a puppy *is* tiring, but you will be glad when she grows into a well-behaved dog.

INTERACTING WITH CLIENTS AND CO-WORKERS

Speaking with Clients

Questions, questions, questions. In a typical day, a veterinary assistant working in a busy animal hospital might answer more than a hundred questions. When you become a veterinary assistant, much of your day will be spent answering clients' questions and giving them directions about pet care. Sometimes, your job will be frustrating. Clients will ask you the very questions that you have just finished explaining in your directions. Other times you may feel clients' questions are irrelevant, unnecessary, or a waste of precious time. But no matter what the question is, you should never appear rushed or irritated. Answer all questions patiently and tactfully.

One way to limit the number of questions is to give clients written instructions along with your verbal instructions.

VETERINARY ASSISTANT: Mrs. Elkin, you're to give Baby one pill, with food, three times a day. Please read these directions, and then feel free to call me if you have any questions.

Some veterinary assistants use the "echo" technique to make sure clients understand directions. To do this, you simply ask the client to repeat your directions back to you (nicely, of course). Then, you listen very carefully to make sure he understands.

> VETERINARY ASSISTANT: Mrs. Elkin, do you have any questions about how many pills Baby gets each day?
>
> CLIENT: You said she gets three a day, with food. Right?

Because you work in an animal hospital, clients will sometimes ask your advice about their pet's medical problems. Be careful not to discuss different types of treatments or their merits with clients. And do not discuss your own personal experiences with a pet's illnesses either. Tactfully refer all concerns to the veterinarian.

> CLIENT: Do you think this lump on Millie's head could be cancer?
>
> VETERINARY ASSISTANT: I'm sorry, Mrs. Maxwell, I really do not know. You should ask the veterinarian to take a look at it.
>
> CLIENT: If it is cancer, do you think there is a cure?
>
> VETERINARY ASSISTANT: I'm sure the doctor will be able to answer all of your questions.

When you are interacting with clients, make sure your feedback is appropriate. That is, the feedback you give should be a true reflection of your concern and understanding of the message. For example, if the veterinary assistant in the previous example snapped her response in a harsh voice, Mrs. Maxwell would sense impatience and annoyance behind the words. However, if the assistant used a pleasant voice and punctuated her words with a comforting smile, Mrs. Maxwell would sense the genuine caring that a veterinary assistant ought to be projecting toward clients. The veterinary assistant shown in Figure 5-4 is projecting a positive, upbeat image to the client.

Listening

Picture this scene: You are a veterinary assistant at the Animal Care Clinic. Mrs. Harris, a client, is leaning over her 12-year-old Irish setter, Daisy, while she waits for the veterinarian to examine her. Mrs. Harris is holding her head in her hands. Concerned, you ask, "Are you okay, Mrs. Harris?" Mrs. Harris looks up, but does not look you in the eye when she answers, "Yes, I'm fine. I'm just a little bit tired." However, when you look at her, you see that she is crying. Her words say she is fine, but the rest of her says she is upset about something. Mrs. Harris is sending mixed messages that put you at a loss for words.

Listening to what others say is an important part of the communication process and an important part of your job as a veterinary assistant. But listening is not complete without observation. Keep in mind that listening involves

FIGURE 5-4 Eye contact and a smile are strong interpersonal communication tools. *(Setting courtesy of Airpark Animal Hospital, Westminster, MD)*

the eyes as well as the ears! Good listeners hear exactly what another person says, and they compare that message with the person's expressions and **body language,** the gestures and mannerisms that help a person communicate.

Good listeners also know when *not* to speak. Have you ever had trouble putting your ideas into words, only to be frustrated all the more by someone else jumping in and filling up the silence during which you were thinking?

HOW TO LISTEN

- Prepare yourself to listen. Stop all other activity and focus your attention on what is being said.

- Look at the speaker.

- Concentrate on what is being said.

- Do not allow distractions of any kind to interfere with your listening.

- Listen with empathy.

- Listen not only to what is being said but also to how something is said.

Use short periods of silence to allow clients time to collect their thoughts. But also look for nonverbal cues that the silence has gone on too long and is causing anxiety. If a person is fidgeting or rubbing his or her temples, it is time to break the silence.

Silence at inappropriate moments can sometimes be a sign of hostility or mental disturbance. Once in a while, you may meet a client who is so upset about his or her pet's illness or death that he will only speak to the veterinarian. In this type of situation, you need to know when not to push things.

In communication, listening is generally more important than talking.
—From The Job Survival Instruction Book

Observing

As you just learned, you need to pay attention to people's body language as well as their words. In order to be a good communicator, you need to be able to recognize nonverbal cues—your own and those of other people. Observing nonverbal behaviors is not always easy. In fact, interpreting nonverbal cues can be complicated.

Often, people do not even realize they are communicating nonverbally. We can control our body language to some degree, but most nonverbal communication is not under conscious control. Many of us have habits—nail biting, finger tapping, hair twisting—that reveal our nervousness or boredom. Nervous habits are easy for an observer to recognize, but many other types of nonverbal cues are not so obvious. In fact, the same gesture or facial expression may mean one of several different things. For example, if you sit with your arms folded over your chest, it could mean you are:

- trying to protect yourself from somebody or something
- hugging yourself as a form of comfort
- conscious about your physical appearance
- cold and trying to warm up

Also, keep in mind that nonverbal behaviors do not mean the same thing to everyone. For instance, in North America a nod of the head means yes and a shake of the head means no. But in some cultures, a nod means no and a shake means yes. A burp at the dinner table is considered rude in some cultures but a compliment in others.

Therefore, you should not jump to any conclusions about the meaning of any particular nonverbal communication. If you are dealing with a client and wish to understand a certain nonverbal message, you might discuss your observations with him. Then ask the client about his or her feelings. Just remember that your observation should never be rude or judgmental. Use tact, which as you remember, means doing and saying the right things at the right time. Do not play the amateur psychiatrist or invade the client's privacy.

Rude: You do not have to cross your arms over your chest, Ms. Summerfall. The veterinarian will not hurt your puppy.

Tactful: Ms. Summerfall, I would be happy to turn up the heat if you are cold.

Observing and interpreting nonverbal behavior is especially important when the individual's body language contradicts his or her words. In these cases, the individual may be expressing through nonverbal communication what he is unwilling or afraid to say out loud. For instance, observe the non-verbal behavior in Figure 5-5. What is the client's body language saying? How does the nonverbal communication contradict the client's words? Despite his words to the contrary, it is clear from his body language that the client in Figure 5-5 is reluctant to hand his kitten over to the veterinary assistant, even though he says he will. In this instance, the veterinary assistant should try to reassure the client that his kitten will be just fine with her. Her words and her actions— showing the kitten some extra affection—will most likely ease the client's fears.

Paraphrasing

Really good listeners sometimes **paraphrase** what they hear.

Paraphrasing, or using different words to express the same idea, is a great way for veterinary assistants to develop their listening skills. When you para-

FIGURE 5-5   Your body language can communicate feelings that you do not want to express in words.

phrase, you listen to a speaker and then repeat his or her message in your own words without changing the meaning.

> SPEAKER: I work so hard all year long. It does not seem that I would be out of line to expect decent accommodations and good weather for two lousy weeks!
>
> LISTENER: When you finally have a vacation, you want things to go well.

Repeating the speaker's words **verbatim** (word for word), does not have the same effect. Word-for-word repetition could just mean that you, like a parrot, have a knack for repeating what you hear. Besides, clients are apt to get annoyed with you if you are always saying exactly what they say. But saying what they mean is a different matter. When you do this, clients feel that you are really paying attention to them and are fully engaged in the conversation.

In addition, by paraphrasing, you can test whether you heard the message correctly and understand the speaker's intentions. Hearing their own words reflected back by a veterinary assistant also helps clients. People do not always say exactly what they mean. If you missed the point, they can tell by your paraphrasing and then clarify what they really meant to say.

> VETERINARY ASSISTANT: Could you describe Muffin's symptoms for me?
>
> CLIENT: Well, she will not eat, and she's throwing up.
>
> VETERINARY ASSISTANT: So Muffin will not eat, and she vomited this morning?
>
> CLIENT: Well, actually, she threw up yesterday and she hasn't eaten since then.

Paraphrasing can also enhance a client's willingness to talk. Consider the scenario with Mrs. Harris and her 12-year-old Irish Setter. Remember that Mrs. Harris has just told you she's tired, even though she's obviously upset.

> YOU: Mrs. Harris, are you sure you're all right?
>
> MRS. HARRIS: Well, now that you mention it, I am really worried about Daisy. She's supposed to have an operation, and I'm worried she might not make it. You know, she's getting really old.
>
> YOU: I can certainly understand your concern, Mrs. Harris. But try not to worry. The doctor will take very good care of Daisy.

By paraphrasing Mrs. Harris's response, you were able to get her to open up and discuss what was really bothering her—her fear of losing her dog.

Showing Respect

As a veterinary assistant, you will be dealing with clients and co-workers of different cultures and races. Treat people equally, as individuals, no matter what personal feelings you have about their social class, skin color, sexual orientation, physical challenge, or any other difference from yourself.

Occasionally, you might have to communicate with a client from another country. If this is the case, he might have trouble understanding the language and culture of your country. This is one reason why it is so important for veterinary assistants to have empathy, the ability to understand the way others feel, and to respect every human being.

Certain clients with handicaps or disabilities may present the need for improving your professional skill, care, and judgment. It is important to remember in dealing with these clients that although they may *have* a handicap, *they are not a handicap.* They are total persons, just the same as you or anyone else.

You will likely come in contact with clients who are either blind, deaf, or physically disabled. Others might not have certain mental or emotional advantages that most other people have and take for granted. You should never judge, label, or stereotype such clients. If you feel awkward about them, you will make them feel uncomfortable. Treat all clients as people first. You should certainly make allowances for whatever disabilities they have, but you should never identify these clients only by their handicaps or disabilities. If you do, you will be gearing your entire approach to them based on what they lack rather than on what they have.

TIPS FOR TREATING CLIENTS WELL

- Always offer a pleasant greeting when clients come into the practice.

- Treat all clients equally, with respect and honesty.

- Address clients by name—Mrs. Johnson, Ms. Arable, Mr. Loftus, Professor Reiss, Doctor Lenardo. Use first names only for children or adults who ask you to.

- Try to relieve clients' fears regarding their pets' care by carefully explaining all procedures.

- Explain technical terminology when it appears that the client does not understand what you are saying.

- Be sure that the reception area is kept tidy and comfortable for clients. If any patients have an "accident," clean it promptly and cheerfully to prevent the client from becoming too embarrassed.

- Offer assistance if you observe a client experiencing difficulty with a door, chair, or the animal's carrier.

Respecting others requires you to understand your own beliefs and **prejudices**. Most people do not believe themselves to be bigoted or prejudiced. But in reality, every person is (at least a little bit). We all have preconceived opinions and biases. We are all partial to some things and not to others. But it is hard to see our opinions and biases as prejudices because they seem normal to us. Because of this, most people do not even realize how their personal prejudices affect the way they treat others. Waiters and waitresses often place the bill in front of a man without asking who will be paying. People often call a veterinarian "he" without even thinking she might be female. Clients are often expected to be white if their names do not fit ethnic stereotypes.

Stereotypes are preconceived ideas about a group of people made without taking individual difference into account. Stereotyping lumps groups of people together, assigning them the same traits and behaviors, simply because they belong to a certain social group. Stereotyping is a symptom of prejudice. Veterinary assistants who really respect individual differences and who really care about people—all people—will not label individuals. You must be sure that your personal beliefs or prejudices will not interfere with your ability to provide equal, objective professional care to each and every client. If you learn all you can about the cultural groups in your community, you will avoid offending people who have different cultural practices than you do, and you will earn their respect and confidence in the process.

PREJUDICES THAT AFFECT VETERINARY CARE

- Racial or ethnic prejudices
- Gender prejudices (discrimination against women; "man bashing")
- Religious prejudices
- Prejudice against those who do not speak English or those who speak it poorly
- Prejudice against people who have AIDS
- Hidden prejudices:

    dislike of children

    dislike of the elderly

    dislike of overweight people

    dislike of divorced people

    dislike of people on public assistance

    dislike of the mentally or physically challenged

Telephone Manners

It is very easy to forget that the voice on the other end of the telephone line is an actual person, and as such is due the proper attention and respect due to any human being. Here are a few points to remember when talking on the phone:

- *Answering.* When the phone rings, be sure to answer it in a reasonable amount of time. Generally speaking, you should answer before the third ring.

- *Greeting.* Always use an appropriate greeting when answering the phone. You should identify the name of the business, give your name, and then ask how you can help the caller. This ensures the person calling that he has dialed the correct number, explains who he is talking to, and invites him to make his or her request.
- *Listen.* Just as you would with a person you were speaking to face-to-face, you should be sure to listen to what the caller is saying, and answer promptly and courteously. Avoid doing other things while talking on the phone with clients, as they can distract you from the call.
- *Manners.* Use the same manners on the phone that you would use in person. "Please," "thank you," and "you're welcome" show respect for the person on the phone and portray you, and by extension, the practice, as professional.
- *Avoid putting callers on hold.* When a caller is put on hold, he can quickly feel ignored, even though you might be looking up the information he is requesting.
- *Take notes.* Keep a pad by the telephone to take notes when someone calls. Write down the client's name when given, and if he is requesting information, make a note of what he is looking for. This helps reduce the number of times you need to ask the person to repeat the information, and shows that you are paying attention to the call.
- *Know why you are calling.* If you are the person placing the call, know the reason you are making the call and what you hope to gain from it. Be sure to have any information or files that you might need on hand while making the call to avoid delays. You do not want to waste the time of the person you are calling any more than you would want a caller to waste your time.

Your Body Language

You already know that you need to observe your clients' nonverbal messages in order to understand what they are feeling. But, what about your own nonverbal communication? Surely, you will send out messages to clients as well. You need to be aware of your nonverbal messages, and whenever possible, make sure they are positive. *The way you dress, your facial expressions, your eye contact, and your body movements all say something about you*

Earlier, you learned how important it is to convey a professional image. When you take pride in your appearance, you send clients a message that you are a professional who likes things done right. Always practice good hygiene, groom yourself attractively, and dress appropriately (Figure 5-6).

Because veterinary assistants work with animals, they usually dress casually. Keep in mind, though, that first impressions are important, so "casual" should not mean "sloppy." Casual clothes should be neat, clean, and in good taste.

Another obvious positive nonverbal message is a smile. Veterinary assistants should be quick to smile when interacting with clients. A comforting,

FIGURE 5-6 Your appearance is the first message you send to a client.

sincere smile can go a long way toward relieving anxieties. However, be careful not to force a smile; you do not want clients to feel you are faking your hospitality. And, when a client is deeply troubled, be especially careful to smile only at appropriate moments so that you will not be perceived as insensitive or flippant. Your facial expression should always be appropriate to the tone of the message you are giving or receiving. But do not frown! If you look glum, you will only add to the nervousness of clients whose animals are sick or in pain.

GOOD BODY LANGUAGE FOR THE VETERINARY OFFICE ASSISTANT

- Face the client and hold your gaze steady.
- Hold your arms at your sides or gently fold them.
- Stand straight with your legs upright or gently crossed.
- Keep your posture erect but not rigid.
- Stay approximately one arm's length away from the client.
- Always wear professional attire.
- Always practice good hygiene.
- Keep a relaxed facial expression or match the expression of the client.
- Speak in a moderate and clear tone of voice.
- Wear a name tag

Eye contact is another important nonverbal communication. When you look someone in the eye, in effect, you are saying, "I am interested in speaking with you and hearing what you have to say." Do not look away from people when they are talking to you, or they will feel that you are not interested in them. But do not stare intently, either, or you will communicate the opposite—too intense interest, or even hostility.

If you make eye contact when you speak, clients will sense that you are open and honest. If you do not make eye contact with clients, they may perceive you as shifty and untrustworthy. Eye contact gives you a good view of what is going on inside people's heads. You can usually tell whether clients have understood your messages by the look in their eyes.

Your gestures (body movements) can enhance your interactions with clients, too. A welcoming or parting handshake is a powerful gesture that is a sign of friendship (Figure 5-7). And a comforting touch on the shoulder or a pat on the back at an appropriate moment can make a client feel secure. However, some people do not like to be touched by strangers, so do not approach every client in the same way. The touch has to be a genuine, friendly gesture attached to a moment of closeness. For instance, you might put your hand on the shoulder of a child whose dog is sick or hurt.

Handling Angry Clients

On occasion, a client will become angry. More often than not, this anger is directed at the veterinary assistant. Frequently, the attack comes so suddenly that you feel shocked and frustrated. The situation will be easier to deal with if you realize that the person is in a highly emotional state.

Unfortunately, there is no magical formula for calming people down. If you know the person very well, you might be able to figure out the source of

FIGURE 5-7   A handshake is a sign of friendship.

anger and then deal with it. (The real source of anger may be different from whatever subject the person is yelling about.) However, you will seldom know clients well enough to do anything except listen.

Keep a positive attitude; that is, try to respond to people, situations, or objects in positive rather than negative ways. That goes for your nonverbal messages too—make them positive. Remember, a frown shows disapproval rather than empathy. The following list indicates other negative nonverbal behaviors:

- A mumbling tone of voice shows lack of respect.
- Fidgeting or turning away from the client shows a lack of warmth or a desire to escape.
- Evasive eye contact indicates insincerity.
- Pointing a finger, shaking your fist, or speaking in a loud voice sends a confrontational message.

Instead of these negative nonverbal behaviors, use positive head nods, devote your full attention to the speaker, smile, make appropriate physical contact, be honest and look the speaker in the eye, and use a natural tone of voice.

Here are some additional tips to help you cope with an angry client:

- When you notice that a client is becoming angry, show him into a private area such as an examining room, an office, or some other less-public spot.
- Permit the client to tell you about the problem. Remain objective. Listen attentively.
- At no time should you place any blame for the problem. Do not say anything critical of the veterinarian or anyone else on the staff. Do not even say anything the client might interpret as critical.
- Keep yourself under control, no matter what the angry client says. At all costs, avoid arguing!
- Apologize for any misunderstanding.
- Assure the client that the matter will be resolved and that any conflict will be avoided in the future.

Angry people often say things they are sorry for later, so do not take personal offense at anything an angry client says. If the client raises a reasonable objection to an error made by someone in the office, do not get defensive and try to justify the error. The client's complaint will be handled by the veterinarian, so you should say nothing to indicate that the conflict will be resolved in any certain way. If you treat an angry client with tact and understanding, you could save the veterinarian's reputation from the damage a disgruntled client can cause.

Anytime you argue with a customer, you lose. Even if you win, you lose.
—From The Job Survival Instruction Book

VETERINARY TEAM COMMUNICATION

By now, you know you need good interpersonal communication skills to communicate with clients. You also need to use your interpersonal communication skills when you interact with the veterinarian and your co-workers. The veterinary staff is a team working together. Although you will probably be assigned specific duties, keep in mind that you are a team member, so you need to be flexible. *"That is not my job" is one expression that you should eliminate from your vocabulary.* But you also must be careful not to overstep your boundaries. There is a difference between pitching in and taking over.

As you can see from the employee evaluation form in Figure 5-8, the ability to cooperate and get along with others is very important to employers. You need to be cooperative, dependable, polite, and patient with your co-workers, even though your co-workers may differ from you in their personality traits, beliefs, values, and work habits. You must be above pettiness and moodiness. Everyone gets down in the dumps or irritated sometimes, but the competent veterinary assistant does not bring a bad mood into the office. Never let your personal problems interfere with your professional interactions, even when your problem is with a co-worker. Put the objectives of your job before your personal feelings.

The veterinarian is the leader of the veterinary team. If you work for a small private practice, you may be the only assistant and perform both administrative and clinical duties. In this case, a good relationship with your employer is even more important than in larger veterinary clinics and animal hospitals. But regardless of where you work, a good working relationship with the veterinarian is always important to a winning veterinary team.

> *Employees who are helpful and easy to get along with are valued more than difficult people with better skills.*
> —*From* The Job Survival Instruction Book

Your attitude toward the veterinarian must be respectful. Always call the veterinarian by his or her professional name.

VETERINARY ASSISTANT: Good morning, Dr. Masterson. You have several messages on your desk.

VETERINARY ASSISTANT: Please make yourself comfortable, Mr. Levine. Dr. Masterson will be with you in a few moments.

The veterinarians you work for are entitled to your loyalty. They are trained and generally highly regarded professionals working in an often physically and emotionally stressful position. Although you might not agree with a particular behavior, action, or outcome, you should be careful not to assume the cause is a lack of skill, knowledge, reliability, or professionalism. Veterinarians generally work to the best of their abilities, with the best interests of the client and patient at heart. This does not mean that you or the clients will always

EMPLOYEE EVALUATION

Employee's Name   Position/Title   Date of Review

Department   Supervisor   Date Employed

Type of Review: _____ Probationary _____ Six Month _____ Annual

Date of Last Evaluation _____

Total days absent _____   Total days late _____

Quantity of Work: inferior / careless / does just enough / average / above average / exceptional

Volume and consistency

Quality of Work: very poor / fair / good / acceptable / excellent / superior

Accuracy and neatness

Initiative: lacking / needs pushing / adequate / excellent / superior

Motivation

Judgment: poor / unreliable / limited / plans well / reliable / superior

Planning work, making decisions

Adaptability: poor / slow / satisfactory / good / excellent / superior

Adjusts to change

Cooperation: uncooperative / difficult / cooperative / excellent / superior

Getting along with others

Speed: slow / moderate / average / above average / superior

Rate of work

Job Knowledge: very little / limited / adequate / average / good / superior

Since last evaluation, employee has   Recommended for pay increase

_____ Improved _____ No change _____ Regressed   _____ Yes _____ No

Overall impression of this employee:

_____ Unsatisfactory _____ Poor _____ Fair _____ Good _____ Excellent _____ Exceptional

Comments:

_____
Supervisor's Signature

FIGURE 5-8 Employee evaluation form.

agree with a veterinarian's recommendations. However, even if you feel strongly that the veterinarian has done something wrong, you should not express this to clients, co-workers, or any member of the general public. Never say or do anything that would cast an unfavorable light on his or her reputation. If you are experiencing a problem with a veterinarian, speak to your supervisor in private.

DISCRIMINATION

In the section, "Showing Respect," you learned about discrimination in regard to your clients. You learned how to recognize your own prejudices and insensitivities. But what do you do if the discrimination is directed toward you? By law, employers are not allowed to discriminate based on race, color, national origin, religion, sex, family status, handicaps, or age. If you feel you have been denied a job or have lost a job for any of these reasons, contact the Equal Employment Opportunity Commission (EEOC) or the Canadian Labour Relations Board. However, if you are the victim of more subtle discrimination—say, a co-worker makes offensive comments about your weight, or you feel a client is harassing you—then you will have to use interpersonal skills to handle the situation.

GAME RULES FOR VETERINARY TEAM MEMBERS

- Be sensitive to the people you work with.
- Make adjustments to cooperate with fellow employees.
- Show interest.
- Do not take personal problems into the office.
- Express appreciation to teammates.
- Be courteous.
- Be open to new ideas and concepts.
- Keep communication lines open between all staff members.
- Be honest with yourself and others.
- Keep the private business revealed to you by co-workers to yourself.
- Do not spread rumors—if you have a criticism of an employee, take it directly to that person.
- Do not gossip about clients, veterinarians, or other personnel to anyone—not even your spouse or your best friend.
- If misunderstandings occur, clear them up immediately.
- Admit your mistakes.
- Accept constructive criticism graciously and with an open mind.

VETERINARY ASSISTANT: Mary, thank you for your concern, but I have heard enough comments about diet programs and exercise videos.

VETERINARY ASSISTANT: Mr. Stern, as I have told you, I simply do not date clients. Please do not make any more personal advances.

In some situations, confronting the discrimination may be risky business. What if your employer, an employer's relative, or a favored assistant is doing the discriminating? In these instances, reporting the situation may result in your termination or even in a messy legal proceeding. You will have to judge for yourself whether to turn your back, resign, ask for a transfer, or confront the situation. Often, you can help correct a situation by subtly hinting that you are uncomfortable working in an environment in which unfair practices are going on. Employers may be sued by employees who are harassed.

IMPROVING YOUR SPEECH

A person's speech can interfere with his or her message. For instance, if a person speaks too quickly and mumbles, you might miss some of what he said. And, if a person speaks too slowly and uses poor grammar, you might lose interest.

When you speak to clients and co-workers, try to speak at a moderate rate. In addition, try to speak clearly and effectively. Always use direct, concise language that the client can understand.

Also be sure to use correct grammar and pronunciation, and to speak in a pleasant tone of voice. Sound tough? It is really not difficult if you practice. We will begin by analyzing the speed of your voice to see if it needs improvement.

How quickly or slowly do you speak? Do people often ask you to repeat a comment? Do they seem uninterested in what you say? Your voice should sound "natural." If you have a natural tendency to be a fast talker, slow down. If you speak too slowly, speed up. An average rate of speech should be approximately 120 words per minute. You can measure your rate of speed by reading the passage in Figure 5-9 out loud, taking time to pause where you would if you were engaged in a conversation. Read the passage through silently once or twice to familiarize yourself with the words. Time yourself. When you finish, divide 600 by the number of minutes it took you to read. Do not round off to the nearest minute, but to make your math simple it is all right to round off to the nearest 15 seconds. For instance, if it takes you 5 minutes and 15 seconds, divide 600 by 5.25. If it takes you 5 and a half minutes, divide by 5.5. If it takes you 5 minutes and 45 seconds, divide by 5.75. A speaker with an average rate of speech will take approximately 5 minutes.

So, how did you do? Do not worry if the rate at which you speak was less than perfect. Remember the old adage, "practice makes perfect."

COMMUNICATION BARRIERS

While there are many barriers to effective communication, Thomas Gordon, an expert on interpersonal communication, has identified 12 of the most common ones. These conversation stoppers are almost guaranteed to block the flow of communication between individuals, and can even end friendships! How many do you recognize?

Criticizing. Making a negative evaluation of the other person's actions or attitudes. "You brought it on yourself; you've got nobody else to blame for the mess you're in," or "Can't you do anything right?"

Name calling. Putting down or stereotyping the other person. "You hardhats are all alike," or "What a dope!" or "Just like a woman," or "You're really dumb."

Diagnosing. Analyzing why a person's behaving a certain way; playing amateur psychiatrist. "You're just doing that to irritate me," or "I know just what's wrong with you," or "Just because you went to college, you think you're better than I am."

Praising evaluatively. Being too nice by saying things about a person that are excessive or aren't really true. "You're perfect," or "You're the best typist in the world," or "I've never seen anything like that report, really fabulous."

Ordering. Commanding the other person to do what you want to have done. "I want you to do this report right now. Why? Because I said so!" or "Get these letters out right now and take your break later."

Threatening. Attempting to control the actions of others by warning of negative consequences. "If we don't get along better, I'm going to tell Mr. Smith about you," or "You'll finish that report tonight or else!"   or "Just come in late again and see what happens."

Moralizing. Telling another person what to do or preaching what you believe is right or proper. "You shouldn't get a divorce; think about what will happen to the children," or "You ought to tell him you're sorry," or "You can do much better than that if you try."

Bully questioning. Asking questions that are often conversation stoppers because the response must be a forced yes or no. "Are you sorry you did it?" or "Well, weren't you supposed to know that before you attended the meeting?"or "You mean you didn't take the report with you?"

Unwelcome advising. Giving the person a solution to a problem even when the person didn't ask for one. "If I were you, I'd sure tell her off!" or "That's an easy one to solve; first you . . . ," or "What you need to do is go to night school."

Diverting attention. Pushing the other person's problems aside through distraction. "Don't dwell on it, Sarah; let's talk about something more pleasant," or "You think you've got it bad—let me tell you what happened to me!"

Logical argumentation prematurely. Attempting to convince the other person with an appeal to facts or logic without knowing the factors involved. "Look at the facts: if you hadn't left work early the other afternoon, we would have finished the report, and Ms. Smith wouldn't be upset," or "By devoting 20 minutes to opening the mail in the morning and concentrating on getting all your typing done before lunch, you should be able to spend every afternoon changing the files over."

False reassuring. Trying to stop the other person from feeling negative emotions. "Don't worry, it's always darkest before the dawn," or "It will all work out okay in the end," or "There's no point in crying over something that you can't do anything about."

FIGURE 5-9   Use this passage of 600 words to check your rate of speaking.

Volume, Pitch, and Tone

Your speaking **volume** is the degree of loudness. The **pitch** of your voice is its highness or lowness of sound. **Tone** communicates mood or feeling; your voice can sound soft, rough, sweet, harsh, excited, or bored. The volume, pitch, and tone of your voice will vary according to circumstances. Listen to someone who is thrilled about something. That person's voice will have a high, louder-than-usual quality to it. Or, listen to someone giving a speech over a microphone; the tone will usually be lower and richer.

Some people speak so loudly that they blast the listener's eardrums. And others speak so softly that they can hardly be heard. It is difficult to concentrate on either type of voice. Of course, there are times when shouting and whispering are the appropriate speaking volumes. But do you shout or whisper when you speak in normal conversation?

Although it is good to maintain a moderate volume, pitch, and tone in the office, be careful not to take moderation too far. Speaking in a **monotone** voice—one that does not show a change in feeling or pitch—is a quick way to put your listener to sleep. A voice with variety is more pleasant to hear than a constant humming sound. Raise and lower your voice as you speak. This variety makes you appear more interesting and, therefore, people are more likely to listen to what you have to say. Use a pleasant tone of voice that shows enthusiasm and warmth. Remember that your voice represents your personality. Your speech should match the smile on your face.

Did you ever notice that some people change their voices when they are on the telephone? The phone can bring out the worst in people's speech. Some people who speak at a moderate, level volume in face-to-face conversation will use the telephone like a bullhorn. Others speak as if the phone wires amplified their voices. Since a significant part of the veterinary assistant's job includes answering the phone and making calls, you should have a friend critique your telephone use. Arrange with someone to receive your call and place one to you. Have your friend note your volume, pitch, and tone. (We present more about telephone management in Chapter 7.)

Enunciation and Pronunciation

Do you remember Eliza Doolittle in *My Fair Lady* repeating over and over, "The rain in Spain stays mainly on the plain"? She finally got it! She learned enunciation and pronunciation. The way you form or articulate your words is called **enunciation**. An example of poor enunciation is slurred speech. When Eliza Doolittle learned proper enunciation, she pronounced words correctly and spoke them distinctly.

To enunciate clearly, you must use your lips, teeth, jaw, and tongue to form precise sounds. To practice your enunciation, read out loud into a tape recorder, concentrating on each word, and play back your reading to hear how you sound. Avoid the following common mistakes:

Do not sound a silent *h*.

- heir
- honor
- heiress
- honest
- honorable

Be sure to sound the *h* in each of these words.

- wharf
- when
- where
- which
- while
- whip
- whiz
- why

Distinguish between the sound of *ern* and the sound of *ren*.

- south*ern*
- west*ern*
- north*ern*
- east*ern*
- child*ren*
- breth*ren*

Do not confuse *pre* with *per*.

- *per*form
- *per*sist
- *per*haps
- *pre*tend
- *pre*vent
- *pre*scription

Sound the final *g*, but do not hang onto it, and do not make it hard like the *g* in *grunt*.

- sitting
- playing
- dancing
- sing
- ring
- thing

Do not run words together.

- Give me (not *gimme*)
- Saw her (not *saw r*)

- Let me (not *lemme*)
- Catch them (not *ketch em*)
- Don't you (not *don't cha*)

TONGUE TWISTERS

If you practice each of the following tongue twisters, you will improve your pronunciation. Try modulating your pitch and tone in an interesting way.

1. Are our oars here?

2. Bring me some ice, not some mice.

3. Suddenly seaward swept the squall.

4. He saws six, long, slim, slender saplings.

5. Amos Ames, the amiable aeronaut, aided in aerial enterprise at the age of eighty-eight.

6. Six thick thistle sticks, six thick thistles stick.

7. A big black bug bit a big black bear.

8. He rejoiceth, approacheth, accepteth, ceaseth.

9. Geese cackle, cattle low, crows caw, cocks crow.

Additional Tips

You might enjoy taking speech, oral communication, or acting classes to improve your oral communication skills. But you can also become more articulate simply by keeping your ears and eyes attuned to language. Here are some suggestions:

- *Listen attentively to those who speak correct, effective English.* Pattern your speech after theirs. You might ask a friend who speaks well to call your attention to any errors you make. Avoid the use of double negatives.
- *To train your ear, imitate a favorite radio or television announcer.* Be sure to choose an announcer with perfect enunciation. Listen to this announcer whenever possible, paying close attention to his or her speech patterns. Imitating this speech pattern, speak into a tape recorder and play it back so you can analyze your progress.
- *Listen to recordings of popular books.* Tape recordings of books are narrated by excellent speakers, and most books are written in standard English. Be aware that dialogue in fiction is often written in colloquial or local dialect. Picking out the differences between standard English and the various dialects is a good way to improve your language skills.
- *Acquire the dictionary habit.* You will hear, and discover in your reading, many unfamiliar words. Do not let new words pass you by. Look them

up in the dictionary. Note their spelling, pronunciation, and meaning. Whenever there is a suitable occasion, use them so that they become an active part of your daily vocabulary.

SUMMARY

Effective interpersonal communication is an important aspect of the veterinary assistant's job. You have to know how to get your message across clearly and fittingly in a variety of situations. Proper communication involves listening, speaking, and appropriate body language. You must always be respectful when speaking with clients, in person and on the telephone, even when they are angry. Interpersonal communication skills are also vital to your interactions with the veterinary team. Attentive, respectful interactions make for a stronger and more enjoyable working environment.

The way you speak is often as important as what you are saying. You should always speak clearly, with proper enunciation and pronunciation, and ensure that the tone, pitch, and volume of your voice match the message you are trying to convey.

CASE STUDY

Mrs. Simms arrived for her cat's appointment with her two rambunctious boys in tow. She checked in with Walter, the veterinary assistant watching over the reception area, then took a seat with the cat, Button, in her lap while the boys chased each other around the room. Their mother quietly tried telling them to sit still, obviously embarrassed by their behavior, and clutching Button a bit too tightly as she spoke to them.

Walter was not very fond of young children, and wanted to yell at them and tell them to quiet down, but he knew it was not his place. He was also concerned about the way Mrs. Simms was holding her cat, which was now starting to fidget and mew loudly.

Pulling out a box from behind the counter, Walter asked, "Excuse me, Mrs. Simms? Do you think the boys would like to color? We have some crayons and animal safety coloring books that I am sure they would enjoy."

Accepting his offer, Mrs. Simms came up to the counter and handed each child a coloring book and some crayons. The boys quickly took them back to their seats and began coloring enthusiastically, but quietly. Walter also offered her the use of a leash to hook on the collar of the cat. "This will let Button walk around without straying too far from you, and will also free up your hands so you can read a magazine while you wait for the doctor."

Mrs. Simms thanked him again and hooked the leash to Button's collar, then sat and read while the cat climbed up to sit on the window sill in the sun.

- How did Walter exhibit the traits for positive interpersonal relations and use them to diffuse the situation?

- How and when did he deal with his own prejudice?
- What might have happened if Walter had not intervened?

REVIEW

Match each of the six terms related to the communication process with its correct definition on the right.

1. message

2. sender

3. channel

4. receiver

5. feedback

6. reference points

a. form of communication such as writing or speaking

b. characteristics such as education and experience that determine how someone expresses and understands messages

c. the idea that one person wants to get across to another person

d. a return message

e. the person who first initiates a message

f. the person listening to a message

Select the best, most appropriate response for the veterinary assistant to say in each situation.

7a. We all forget our checkbooks once in a while, sir, but payment is still due on the day services are rendered.

7b. Do not worry about forgetting your checkbook, Mr. Jones. You can drop off the check later or put it in the mail today. I can give you a self-addressed, stamped envelope if you like.

8a. I am sorry to hear about your kitty getting run over, Mrs. Simms. You must be heartbroken. Cats are always running off to do their own thing. I am a dog person myself.

8b. I am sorry to hear about Smudge, Mrs. Simms. I know it is heartbreaking to lose a pet.

9a. We would appreciate it if you would not smoke, Mr. Linder. A few of our clients have breathing problems.

9b. Of course we do not have an ashtray, Mr. Linder. Didn't you see the No Smoking sign?

Indicate whether the following statements are true or false.

10. Sitting with your arms crossed over your chest always means you are cold and trying to warm up.

11. To paraphrase, you repeat what someone said, word for word.

12. To provide the best possible care for the client and his or her pet, the veterinary assistant should attempt to determine the message behind nonverbal communication.

13. Body language always reinforces or agrees with the spoken message.

Paraphrase each of the following client's statements.

14. Client: Rusty's sick. He will not eat, and he sleeps too much.

15. Client: My kitten, Flossy, has a cold. Her nose is running and her eyes are all watery.

Fill in the blanks with the correct answer.

16. The gestures and mannerisms a person uses to communicate are called _____.

17. When dealing with an angry client, you should try to keep a _____ attitude.

18. Preconceived ideas about a group of people made without taking individual differences into account are called _____.

Consider the following scenario: A client is shouting at Rhonda, a veterinary assistant, because he has been waiting 45 minutes to see the veterinarian. Rhonda is stunned by the man's anger, and shouts back, "Well, it is not my fault. Stop screaming at me."

19. What did Rhonda do wrong?

20. List six barriers to effective communication.

21. A co-worker mentions that you "preform" well under pressure. What is wrong with her pronunciation of the word perform?

ON-LINE RESOURCES

Business Etiquette (from Careers-Portal)

This article discusses the importance of proper business etiquette in the workplace.

<http://www.careers-portal.co.uk>
Search Terms: Etiquette; Business

Communicating in the Culturally Diverse Workplace (from Jobweb)

This article addresses the issue of cultural diversity in the workplace, including different styles of communication and how to implement effective channels of intercultural communication.

<http://www.jobweb.com>
Search Terms: Resources; Library; Workplace; Culture; Communications

Merriam-Webster On-line Dictionary

Merriam-Webster offers this on-line version of its popular dictionary, including audio pronunciations to help visitors hear and understand how the word should sound.

<http://www.webster.com>

CHAPTER 6

Stress

OBJECTIVES

When you complete this chapter, you should be able to:

- outline the causes of stress and the psychological defense mechanisms used to cope with stress
- list five positive ways to cope with stress

- explain how time management can reduce stress
- list five time management tools

KEY TERMS

stress	repression	malingering
psychosomatic illness	displacement	denial
National Institute for Occupational Safety and Health (NIOSH)	projection	regression
	rationalization	prioritize
	intellectualization	time tracking
burnout	sublimation	time audit
defense mechanisms	temporary withdrawal	

INTRODUCTION

Suppose you are a veterinary assistant in a busy animal hospital. Three of your co-workers are out sick with the flu, and two of the veterinarians are running a half an hour behind schedule. You already have two calls on hold, when you receive an emergency call. Ms. Harris's dog, Fido, has just been hit by a car. Fido is bleeding, and Ms. Harris is hysterical. You are just about to instruct her on how to stop the bleeding when Mrs. Lee's German shepherd attacks Mr. Filbin's cat in the waiting room. The cat scratches Mr. Filbin and gets loose. Two children are screaming.

How do you think you would feel? *Stressed out!*

Unfortunately, when you start your new job, you might find yourself in a similar situation one day. Being able to cope with stress is an important part of a veterinary assistant's job.

WHAT IS STRESS?

Stress is the physical and/or psychological changes that occur in your body as a result of a change in your environment. Stress can be mental or physical tension that results from frightening, exciting, challenging, confusing, endangering, or irritating circumstances—good or bad. Stress may cause physical reactions such as a rise in blood pressure, a sudden rash, indigestion, stomach pain, constipation or diarrhea, frequent colds and flu, headaches, shortness of breath, or a nervous tic. It may cause emotional reactions such as crying, shouting, drinking, sleeping, and becoming angry. Many diseases have stress as a root cause or an aggravating factor. Stress can lead to depression or **psychosomatic illnesses** (real physical symptoms resulting from an emotional or mental condition). As a veterinary assistant, you must cope with stress from all directions—your clients' stress, your co-workers' stress, and your patients' stress, as well as your own stress.

Not all stress is bad for you. Many people perform job duties and other tasks at a higher level when they feel some stress. "Good" stress gives you the energy and desire to reach for goals and solve problems. A deadline can be a positive source of stress. Then again, if it is an impossible deadline, the stress can be too much for you. The stress of reaching for a goal is positive when the mission is accomplished. You feel good when you have finished something. But failure to reach a goal can result in feelings of worthlessness and depression. It is important to have realistic expectations for yourself. Set goals and time frames that are within reach.

CAUSES OF STRESS ON THE JOB

According to "Stress . . . At Work," a report issued by the **National Institute for Occupational Safety and Health** (NIOSH), the federal agency responsible for conducting research and making recommendations for the prevention of work-related illness and injury, the following job conditions can lead to stress:

- *Design of tasks*, including too much work, too few breaks, working too long, and too much emphasis on tasks that have little purpose and underutilize workers' skills and abilities
- *Management style*, including supervisors ignoring the input of workers when making decisions, poor communication, and a lack of family-friendly policies
- *Interpersonal relationships*, including a lack of cooperation and support among co-workers and managers
- *Work roles*, including unrealistic or unclear job expectations, and giving an individual more responsibility than can be reasonably handled
- *Career concerns*, including a lack of job security, little opportunity for professional advancement, and sudden changes without providing time for workers to properly prepare
- *Environmental conditions*, including unpleasant or dangerous physical conditions

Animal hospitals are not the easiest places to work. As a veterinary assistant, you will have to interact with clients, co-workers, and animals of all different personality types. And some of these personality types will be difficult to deal with. Animals do not get sick on schedule either. In fact, illnesses and emergencies always seem to happen at the most inopportune times—whenever you are behind schedule or the veterinarian is not available.

In order to cope with your job, you need to learn to be flexible. In addition, you must learn the art of taking your job one minute at a time, not worrying too much about keeping everything under control. Realize that there are some things you can control and other things that you cannot. Recognizing the difference between the two is an important step in learning to cope with stress.

BURNOUT

Burnout is a term used to describe a condition that results from too much stress over an extended period. Among veterinary assistants, burnout can result from feeling overwhelmed or becoming too involved with clients and their pets. Sometimes, your good work is rewarded by a client complimenting the veterinarian instead of you.

People who experience burnout generally feel a lack of power over their circumstances. They feel that no matter what they do, they cannot change things. The problem is chronic; that is, it will not go away. In attempting to cope, these people withdraw from interactions with others. They experience fear, anxiety, depression, and decreased energy and productivity.

COPING WITH STRESS

Coping Strategies

You do not want to burn out. Instead, you want to enjoy your new career. Fortunately, you can avoid burnout by getting rest, eating right, and exercising (Figure 6-1). The following are some coping strategies that really work:

- *Exercise regularly.* Daily physical activity is a good release for stress.
- *Maintain good eating habits.* A healthy diet with balanced nutrition helps you to feel better physically, lessening your susceptibility to stress.
- *Deal directly with the source of the stress.* Do not let your feelings build up inside of you, because this can actually amplify the effects of stress.
- *Slow down.* Try to limit mental and physical pressures. You do not want to be the source of your own stress.
- *Relax periodically.* Everyone needs to take a break from time to time. We are all human, and we all become fatigued over time.
- *Join a support group.* Talking with other people experiencing difficulties with stress can be very therapeutic.
- *Discuss problems with a trusted friend.* "Keeping things bottled up" will only lead to further stress.

FIGURE 6-1   Rest, nutrition, and exercise can help you cope with stress.

Work off stress with 20 minutes of physical exercise.
—From *The Job Survival Instruction Book*

Some people use negative coping strategies as a quick fix for too much stress. Naturally, this usually makes things worse. Negative coping strategies include:

- *Calling in sick to "punish" the boss.* This only serves to punish your co-workers, the clients, and yourself.
- *Putting little effort into work.* Rather than reduce your stress level, this strategy is likely to increase it as your supervisor and co-workers will be dissatisfied with your performance, and you will generally end up making more work (and stress) for yourself.
- *Alcohol and/or drug use.* Alcohol and drug abuse can have very serious consequences in your professional, personal, and social life.

Stress and burnout are serious problems that cannot always be avoided. But developing strategies for coping with them can be effective in reducing their levels.

Defense Mechanisms

Psychologists and psychiatrists have identified a number of **defense mechanisms**, adjustments we make in our behavior, usually unconsciously, to help us deal with the experiences or feelings that cause us psychological stress. Although you should never act the role of amateur psychiatrist, recognizing and understanding defense mechanisms can help you deal with your clients and co-workers successfully. You certainly do not have to memorize all of the different defense mechanisms in this section, but you might want to refer to the following every now and then, especially when you think you detect someone using one of them.

- *Repression.* Socially unacceptable or painful desires or impulses are pushed out of the conscious mind into the unconscious, without our being aware of it. These feelings may crop up in dreams or in subtle behaviors. For instance, Jane has forgotten her painful memories of being beaten as a child, but she winces with pain every time her husband, Fred, clears his throat the way her abusive father used to.
- *Displacement.* Emotions about one person, idea, or situation are transferred to another, more acceptable or easier target. Fred hates having to work for a woman supervisor, but he cannot do anything about it at work, so he is a tyrant with his wife and children at home.
- *Projection.* One's own ideas, feelings, or attitudes are attributed to someone else. For example, you convince yourself that someone else is to blame so you will not have to take responsibility and feel guilty. Fred does not feel he is a tyrant with his wife and children; it is Jane's fault

---

KNOW YOURSELF

Before you can understand others, you must understand yourself. This quiz will help you take a good hard look at your interpersonal skills. Answer each question as it applies to you.

1. Do you find it easy to start a conversation?

2. Are you able to hold up your end of a conversation?

3. Do you ask good questions? (Good questions are usually open ended, requiring detailed answers instead of just yes or no.)

4. Are you able to talk about things other than yourself?

5. Do you listen without interrupting the speaker?

6. Do you use body language when speaking?

7. Do you draw others into a conversation when they are not contributing their share?

8. Do you avoid exaggerating facts when speaking to others? (Tall tales do not count!)

9. Do you remember names of people when introduced?

10. Do you avoid using dialect, bad grammar, slang, cliches, or jargon in professional or formal situations?

11. Do you enjoy learning about people, their interests, hobbies, and ideas?

12. Do you keep others interested in what you are saying?

13. Do you give others an opportunity to express their views?

14. Are you able to discuss controversial matters without getting angry or upset?

15. Do you pay attention to the conversation without having your mind wander?

If you answered yes to at least 10 questions, your interpersonal skills are probably quite good. But try to work on any weak areas so that you can change no answers to yes.

---

that he has to crack down when he gets home, because she is too weak and does not keep the kids under control.

- **Rationalization.** One's actions are attributed to logical reasons. One does not look at the true motives of behavior. Fred has been trying to quit smoking, but after he blows up at his wife and kids he says, "I haven't had a cigarette all day, so it is okay to have one now."

- **Intellectualization.** Again, reasoning is used to avoid the truth, as a way of denying strong feelings that may be socially unacceptable or difficult to accept. Instead of admitting to herself and to others that she is unhappily married, Jane chatters on and on about how well Fred is doing at work, how good he is at coaching Little League, and how much they are all looking forward to going on vacation.

- **Sublimation.** An instinctual desire or impulse is diverted into a socially acceptable activity. Jane wishes she could see her old high school sweetheart, but instead she takes her kids to the beach a lot and reads romance novels.
- **Temporary withdrawal.** Ways are found to avoid dealing with a painful or difficult situation. Jane watches a lot of television. Fred stops at a lot of bars.
- **Malingering.** Deliberately pretending to be sick enables someone to escape a situation that causes anxiety. Fred sends word with Jane and the kids that he cannot go to his supervisor's dinner party tonight because he has a stomachache, and then he orders a pizza with everything on it.
- **Denial.** To avoid accepting and dealing with a traumatic, stressful situation, the person refuses to admit or acknowledge that the situation exists. Jane is in denial that she was abused as a child, and Fred is in denial that he is addicted to cigarettes.
- **Regression.** During times of high stress, we may return to an earlier mental or behavioral level. When Fred's wife was a child, she used to enjoy her grandmother's homemade chocolate fudge with nuts, and now she makes fudge whenever she is depressed.

If, on occasion, you recognize some of these defense mechanisms in yourself or someone else, do not worry too much. These behaviors are the mind's natural way of coping with stress. However, habitual use of defense mechanisms can indicate a need for counseling. Chronic dependence on defense mechanisms can point to interpersonal communication problems that might be solved if they were faced and analyzed.

PROBLEM SOLVING

Here are some problem-solving steps that you can use yourself or pass on to others.

1. What is the problem? To identify the real problem, it helps to list the actual events or examples of behaviors that have contributed to the problem.

2. How can the problem be solved? Research the ways that this problem can be solved. Books, counselors, and self-help groups are all good sources of information.

3. What would the outcome be if you did X, Y, or Z? To think through the possible outcome of various decisions, be logical and creative.

4. Now you can test your alternatives. Begin with what seems to be the best course of action and continue down your list of options until the results are satisfactory.

Time Management

One of the most common sources of stress is not having enough time to do everything that needs to be done. Because it is impossible to add more time to the day, the best strategy is find ways to save time and streamline work processes. The following are some suggestions for improving the way you manage and use your time at work:

- *Reducing clutter.* In a busy practice or hospital, records, files, memos, and other paper can pile up quickly, making it difficult to find what you need when you need it. To fight clutter, keep yourself organized as you go. Return medical records and other files to their proper location immediately after they are used. Set aside a separate area or tray for mail, memos, and other forms of correspondence. Whenever possible, use the computer to reduce the amount of paper used in the office.

- *Prioritizing.* Each day, make a list of what needs to be done, and then **prioritize** the tasks, tending to the most important items first and leaving the less important tasks for later in the day. If you run into difficulties or emergencies throughout the day, the items on your list that do not get done were the least important. This also helps you to avoid procrastination.

- *Time tracking.* When you are working on an involved project, it is easy to become engrossed in your work and not notice how much time has passed. To avoid spending more time than you plan on something, keep a small clock at your workspace and get into the habit of glancing at it periodically to note where you stand. **Time tracking** helps you to keep yourself on the day's schedule.

- *Awareness.* It is important for you to be aware of what is happening in the practice or hospital. Is there something being worked on elsewhere that will be passed along to you to complete or work on? Is there something that is not being tended to that could grow into a problem? Being aware of what is happening in your work environment helps you to prepare for and prevent future timing difficulties.

- *Time audit.* If you try these strategies and they do not work, try keeping track of what you do each day, and how much time you spend doing it. Throughout the day, take a moment to jot down on a notepad what you just did and how long it took you. At the end of the day, review your notes to see how you spent your time. What did you spend time doing that you did not need to do, or should have spent less time doing? This simple form of **time audit** can help you to identify areas in which improvement is needed.

SUMMARY

Working in a veterinary office can be stressful at times, especially when dealing with clients worried about their unhealthy animals. It is important that you recognize the signs of stress and develop effective coping strategies to help you avoid burnout. It is also helpful to be able to recognize defense mechanisms that help people deal with the effects of stress, both in yourself and others. They may be a simple temporary reaction, or possibly a sign of an interpersonal communication problem that needs to be addressed.

Time management can be an effective tool in reducing stress. Using methods such as reducing clutter, prioritizing, time tracking, awareness, and time audits can help you to make better use of your time.

CASE STUDY

Zelda was reviewing and purging old client files. She had just come back from lunch and was hoping to get through the remainder of the files by the end of the day. It was taking much longer than she had anticipated because the day was turning out to be very busy, with three emergency calls and two walk-ins, in addition to the regular workload.

Erin, another veterinary assistant, came over and asked Zelda for the Vinson file. Zelda became very upset, saying "Why can't you ever get anything yourself? I don't have time to do your job, too!"

Erin replied, "I'm sorry, Zelda, but I saw that you were going through the files and I didn't want to mess up your system. Are you okay? I know today's been a little hectic. I could give you a hand if you're getting stressed out."

Zelda replied, "I'm not stressed out! I'm just trying to get this work done and people keep interrupting me!"

"I'm sorry. I'll get out of your way," Erin said.

"No, wait, Erin. I'm the one that's sorry. You're right, it's really busy today and I am feeling stressed. There's so much to do and there are so many people coming in, I just don't know what to do."

"Do all of these files really need to be updated today?" Erin asked.

"I guess not. I was hoping to get them all done, but it's really not that important. I'll just do what I can today, and take care of the rest tomorrow."

"I think that's a good idea, Zelda."

"Thanks. You know, I'm starting to feel a little better already."

- What conditions contributed to Zelda's stress?
- What signs of stress do you think Erin noticed in Zelda?
- What defense mechanisms did Zelda employ to deal with her stress?
- How were coping strategies used to reduce her level of stress?
- How could using time management techniques reduced Zelda's stress level?

REVIEW

Match each of the 10 terms with its correct definition on the right.

1. burnout
2. defense mechanism
3. displacement
4. malingering
5. projection
6. psychosomatic illness
7. rationalization
8. repression
9. stress
10. sublimation

a. pretending to be sick to escape a stressful situation

b. diverting instinctive impulses into socially acceptable activities

c. making logical excuses to avoid the truth

d. attributing one's own feelings to someone else

e. transferring emotion from the real target to another

f. pushing painful or unacceptable thoughts into the unconscious

g. unconscious behavioral adjustment

h. physical symptoms with a psychological root

i. bodily or mental tension

j. withdrawal caused by too much job-related pressure

Indicate whether the following statements are true or false.

11. Not having enough time to do your job can cause stress.

12. Organizing as you go helps to increase clutter.

13. When prioritizing, put the least important tasks at the top of your list and do them first.

14. When doing a time audit, you should list each task you performed during the day by its level of importance.

ON-LINE RESOURCES

How to Master Stress (from MindTools)

Mind Tools presents this informative site, designed to help people identify different types of stress, the causes of stress, and multiple stress-reduction techniques.

<http://www.psychwww.com>
Search Terms: Stress; Reduction

The American Institute of Stress

The organization's official Web site, dedicated to advancing people's understanding of the role of stress in health and illness.

<http://www.stress.org>

Time Management (from Careers-Portal)

This article explains the importance of time management, suggests strategies for managing time better, and gives a short quiz to help determine if a person needs to manage time better.

<http://www.careers-portal.co.uk>
Search Term: Time management

CHAPTER 7

Interacting with Clients

OBJECTIVES

When you complete this chapter, you should be able to:

- describe a pleasing telephone personality and demonstrate how to handle the types of calls commonly received in an animal hospital
- using an appointment book, efficiently schedule a veterinarian's workday
- explain how interpersonal skills apply to client and patient reception

- explain what euthanasia is and when it might be used
- identify common client reservations and concerns about euthanizing pets
- describe common reactions to pet loss
- identify strategies for dealing with grief

KEY TERMS

screening	tickler file	establishing the matrix
call director	wave scheduling	monitoring
station	flow scheduling	euthanasia
call board	fixed office hours	

INTRODUCTION

Whatever your status in the veterinary practice, when you communicate with a client, you are the face of the entire veterinary team to that person, so it is important that you both appear and behave in a professional manner.

You will find yourself interacting with clients for a variety of reasons, including making appointments, greeting people as they enter, answering questions, and addressing other concerns. The veterinary assistant should be seen by clients as a knowledgeable, helpful professional, so you must know how to act and dress the part. When calling or answering a call from a client, you should know what to say, what not to say, and how to carry yourself. When speaking with clients in person in the facility, you need to know how to address them, how to act, and how your appearance affects your interaction. It is exceptionally important to know what to do when discussing especially serious matters, such as preparation for, and reaction to, euthanasia. Knowing how to interact with clients is a big part of knowing your job.

TELEPHONE MANAGEMENT

One day Bill Moynihan's cat limped into the house on three legs, blood all over her. Recognizing that the cat needed a veterinarian's care, Bill called an animal hospital right away. The phone rang seven times before someone answered in a hurried voice, "Hallo, can ya hold?"

Before Bill could reply, he was listening to music. After five minutes of wondering whether he had even reached the right number, he heard a veterinary assistant say, between chews of food, "Can I help ya?"

"Is this the Animal Care Hospital?"

"Yup," replied the veterinary assistant.

"This is Bill Moynihan. My cat is bleeding all over and I think she has a broken leg. I would like to bring her in right now if the doctor can see us."

BE PREPARED

Have a properly prepared area for receiving and making calls. Include

- Pens, pencils, erasers
- Memo pad for taking notes
- Telephone message pads
- Alphabetized, up-to-date telephone index of frequently used numbers
- A specific place for telephone messages (a spindle, a file, or a tray)

"Why didn't you say it was an emergency? I wouldn't have put you on hold! Bring your cat right in," the veterinary assistant said and hung up.

This, of course, is a lesson in how not to treat a client. If you were treated so discourteously, you would strongly consider taking your pet-care business elsewhere.

The Telephone Personality

The telephone is the main line of communication between the veterinarian and his or her clients, and is often where first impressions are made. Therefore, anyone who answers the phone or makes calls must develop a pleasing telephone personality. And the veterinary assistant in our example broke every rule of telephone etiquette. As you read the following seven rules for a pleasing telephone personality, try to determine what the receptionist did wrong.

1. *Answer promptly*. Make sure that the phone is covered at all times. Try to answer on the first or second ring. Avoid letting the phone ring more than three times.
2. *Identify yourself*. Immediately give the name of your animal hospital or your employer, as in, "Good morning, Dr. Gorlock's office." To make the greeting more friendly or personal, add your own name, as in, "Hello, Animal Care Hospital, Kim speaking."
3. *Speak pleasantly*. Avoid extremes. Do not speak too softly or too loudly, too slowly or too rapidly. Pronounce your words distinctly and with expression; that is, do not sound mechanical, like a robot. Speak naturally, in a relaxed, low pitch so that your friendly personality comes through in your voice.
4. *Be courteous*. Besides using polite language—please, thank you, pardon me—give your caller your full attention, or explain why you cannot. For instance, if you are helping a client carry in an injured dog, tell the caller why you are distracted. Never try to talk to a caller and someone else at the same time, whether it is the veterinarian or a client checking in for an appointment. Courtesy also calls for using simple language—not the medical jargon you will be sure to pick up while working in a veterinary office.
5. *Explain interruptions*. If you must leave the telephone to get information or put the caller on hold to answer another line, explain why and give the caller a chance to respond. Sometimes people *cannot* hold, and it is unfair to assume they can. Always thank a caller for waiting.
6. *Use the telephone properly*. Place the receiver firmly against your ear, with the center of the mouthpiece about three-fourths of an inch from your lips (Figure 7-1). Speak directly into the mouthpiece. Never eat, drink, chew gum, smoke, or put anything in your mouth while you talk on the phone.

FIGURE 7-1   This veterinary assistant is holding the phone correctly. When you speak directly into the mouthpiece, the listener will hear you clearly.

7. *Conclude courteously.* Usually the caller is the person responsible for ending the conversation. So, if you call, you should end the conversation. In any case, be polite and allow the caller to hang up first.

A truly pleasing phone personality should please you, your employer, and the client. Do not try to force yourself into a personality that does not feel comfortable. If you do, your discomfort will show through, and the client may feel that you lack sincerity. Treat people as though you value and appreciate them. Keep your main goal in mind: *to provide excellent service to clients and their animals while upholding the good reputation of your employer.*

Do not say anything on a cordless or cellular phone you would not want to hear on the evening news.
—*From* The Job Survival Instruction Book

Now do you know what the veterinary assistant did wrong in the earlier example? First, she let the phone ring seven times, which is at least four times too many. Second, she did not identify the veterinary office or herself, leaving the caller to wonder if he had reached the wrong number. Third, she sounded unpleasant, speaking as if she were in a hurry and saying "Hallo" instead of "Hello," "Yup" instead of "Yes," and "Ya" instead of "You." Fourth, she was discourteous, asking the caller to hold without giving him a chance to respond. Then she failed to thank him for waiting. Fifth, she did not explain why she made him wait five minutes—though he probably assumed she had been eating. And sixth, she ended very discourteously by implying that the caller was responsible for the delay. She did not even say goodbye.

Screening Calls

Your specific approach to **screening** (handling) incoming calls will vary according to your employer's wishes. It is important to find out to what extent the veterinarian wants you to screen his or her phone calls. For instance, one veterinarian might trust your judgment in deciding which matters need immediate attention. Another might want you to present all the details, minus any interpretation. And each type of situation requires a different approach as well. An appointment for a routine checkup is not handled the same way as an appointment for an injured animal.

It is a good idea to work out a pattern for screening calls. A routine will help you obtain the information you need in a minimum amount of time while you maintain a friendly, warm, interested manner (Figure 7-2). And by keeping control of the conversation—politely—you discourage any tendency the client might have to ramble and digress.

First, you always need to know who is calling so that you can screen the call if necessary. Most veterinarians accept calls from other veterinarians and immediate family members, no questions asked. In most offices, you will transfer such calls as soon as possible. If the caller does not volunteer a name, you can say, "May I have your name, please?"

FIGURE 7-2 One way to have a naturally pleasant phone voice is to smile when you talk on the phone. Your voice will carry your smile.

Next, you need to know why the person is calling so that you can take the appropriate action—screen the call, transfer it, take a message, or make an appointment. Whereas it would be impertinent to ask why Dr. Thompson's husband wants to speak to her, it would be irresponsible *not* to ask why a complete stranger wants to. Usually a doctor prefers to accept only emergency or urgent calls from clients. Other calls will be returned when the doctor has time between patients or at the end of the workday.

Some clients call to ask the doctor for information or services that you could just as easily provide. If you do not immediately know why someone is calling, you can waste a good deal of time, as in the following conversation.

"Good afternoon, Dr. Doolittle's office, Liz speaking. Can I help you?"

"This is Tom Green. Let me speak to the doctor, please."

"The doctor is with a client right now."

"Well, I want to talk to her. I am a client too."

"Perhaps if you leave a message, the doctor will call you later."

"I suppose. Tell her I'd like to have an appointment tomorrow morning for Salty, my parrot."

If the veterinary assistant had transferred the call to the veterinarian, it would have been transferred right back so that the caller could make an appointment! Some callers simply want to talk to the veterinarian and give no indication of their purpose. You can usually prompt these people along by saying, "The veterinarian is with a client right now. May I help you?" If they still hesitate, you might add, "Would you like to make an appointment, or would you prefer to leave a message?" The more specific you are in your questions, the more specific the caller will be in answering.

Another good reason to know why someone is calling is that you can prepare yourself for whatever action you will be taking. For example, if you know right away that Mr. Green wants an appointment, you can have your pencil and appointment book ready. If you know that a client is calling to find out when Fifi got her shots, you can immediately pull the file. A well-organized veterinary assistant can handle the average call in less than 3 minutes.

Now you know how to screen a veterinarian's calls, but what do you say when the veterinarian is busy, but not with a patient? Or what if the veterinarian is not in the office during regular office hours? Avoid the following statements, even if they are true.

- She is not in yet.
- He went home for the day.
- She went out to lunch.
- He is playing golf all afternoon.

Instead, simply say, "The doctor is not in at the moment" or "The doctor is unavailable." Then offer to take a message, and say, "May I ask the doctor to call you?" Remember, the veterinarian is entitled to a private life; you should never give clients a list of numbers to call in an effort to find the veterinarian. However, if the problem is urgent, you will try to reach the veterinarian or the designated emergency standby and assure the client of this fact.

Occasionally, you will answer the phone to find out that the caller dialed a wrong number. The caller may seem confused or insist you are the hardware store, even though you clearly identified yourself. To avoid a lengthy and useless conversation, you should politely set the matter straight by saying something like, "I am sorry, but this is the Animal Care Clinic. Are you calling 555-5678?" This serves the double purpose of alerting the caller to the mistake and keeping her from repeating it.

Making Appointments over the Phone

Much of your time on the phone will be spent making appointments. And, as you learned in the last section, developing a routine will help you avoid wasting precious time. Try to develop a pattern for asking questions. Surprisingly, a number of clients do not volunteer such basic information as their name, the reason for their call, or a convenient time for them to see the doctor. It is your job to *politely*, but efficiently, prompt such people along. For instance, when a client does not specify a time, you might ask one of these questions:

- Do you prefer morning or afternoon?
- Could you come tomorrow at three o'clock?
- When would you like to come in?

Of course, some clients present the opposite problem, as in, "I *have* to see the veterinarian *today* at *ten*." If the schedule allows, you should accommodate the client's wishes. However, when the veterinarian is busy, you must make that clear while making the client feel wanted. "I am sorry, the only time Dr. Thompson has available today is 4 o'clock. Can you come in then?"

Sometimes the schedule is booked for a week or more in advance, and clients become upset because they cannot get a prompt appointment. In such cases, you might say that you will be happy to call as soon as you have a cancellation. And, of course, you should fit in emergency appointments as soon as possible, even if it means canceling another appointment or making someone wait.

Taking Messages

Most offices have standard message forms for you to jot down the date, time, name of caller, return telephone number, and message (Figure 7-3). You should also include your own name or initials, especially in a large practice where any of a number of people might answer the phone. Accuracy is important, so to minimize confusion, follow these guidelines:

- Write down every instruction clearly, to leave no room for misunderstanding.
- Read back, word for word, the entire message after you have written it down.

To: *Dr. Morgan*

Date: *5/12*   Time: *12:30*   A.M. / (P.M.)

WHILE YOU WERE OUT

Mrs. Thompson

of: *the Humane Society*

Phone:

555-0000

To: *Dr. Brown*

Date: *5/12*   Time: *9:30*   (A.M.) / P.M.

WHILE YOU WERE OUT

Mr. Richard Jackson

of: *client*

Phone:

555-1111

FIGURE 7-3   Sample messages.

- Establish clearly whether the message must be presented or acted on by a certain time or date.

Consider the "what if" situations that could arise if the message is not acted on in time. This can be a difficult but important step. For example, suppose the veterinarian is out when a client calls describing symptoms that you recognize as dangerous for an animal. What if the veterinarian does not return soon? Should the client call a backup veterinarian? Should you (not the client) try to track down the veterinarian? Part of your job as a veterinary assistant is to make judgment calls about how to handle "what if" situations.

Coordinating Calls

Unless your office has a number of people answering phones, you are likely to face the problem of coordinating two or more calls at once. Most veterinary offices have more than one phone line, and you may find yourself juggling calls while you are greeting incoming clients, accepting payment from others, and pulling or returning files. Handling several lines while performing other duties takes a bit of coordination —mental as well as physical! But you can handle the situation smoothly by mastering the use of your **call director** (Figure 7-4).

A call director is a telephone unit with **stations** that allow several calls to come in at once. The **call board** has buttons that represent the stations in your office. Station lights tell you how to handle your call director. A slowly blinking light accompanied by a telephone ring alerts you to an incoming call.

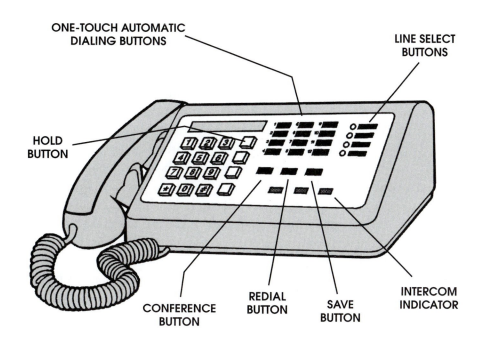

ONE-TOUCH AUTOMATIC
DIALING BUTTONS

LINE SELECT
BUTTONS

HOLD
BUTTON

CONFERENCE
BUTTON

REDIAL
BUTTON

SAVE
BUTTON

INTERCOM
INDICATOR

FIGURE 7-4 Call director.

A constantly blinking light indicates a station on hold. And a steady light indicates a station in use. When a button lights up and the call director rings, you push the station button down *before* you pick up the receiver and answer with your standard greeting. If you do not push the station button first, you could interrupt other calls in progress.

What do you do when you are talking on one station and another call comes in on the call director? Do not panic! The procedure is really easy:

1. Politely ask the person you are talking with to hold. (Give the person a chance to respond in case she cannot hold.)
2. Push down the HOLD button, which is generally red.
3. Push down the station button that is blinking.
4. Answer politely, quickly find out who is calling and why, and then ask the caller to hold.
5. Push down the HOLD button again.
6. Go back to the first station, thank the person for holding, and finish the call as promptly as possible.
7. Return to the second station, thank the caller for holding, and proceed with the call.

To make certain your memory does not fail you, it is a good idea to jot down the facts about any caller on hold.

Of course, this procedure can vary slightly, depending on the circumstances. For example, when the second caller is another doctor or your employer's spouse, you might transfer the call instead of placing the caller on hold. Each office has different policies for transferring calls. You will learn the standard procedure for your office during your training period.

Keep in mind that you might have to change your standard procedure if one of the calls you are juggling will take a long time to handle. For example, suppose that the second caller wants an appointment, and the first caller wants a summary of all the immunizations an allergy-ridden puppy has received in the last year. You will have to locate the first caller's file and spend several more minutes on the phone. In such a case, you may take the first caller's phone number and call back as soon as you have found the information. Or you might ask the first caller to hold while you quickly schedule the second caller's appointment. If, however, you are away from any caller for a length of time, it is polite to return every so often with, "Mr. Jones, I am checking your records" or "I will be with you in a moment."

Sometimes more than one station rings at once. When that happens, follow the same procedure outlined for two calls. Simply ask each new caller to hold while you promptly complete the one in progress. You should generally try to handle calls on a first-come, first-served basis, but urgency or time considerations could justify a change in order. And it is better to finish one conversation at a time than to switch back and forth between lines. That is annoying for the callers and difficult for you.

HOW TO HANDLE COMMON CALLS

If the caller wants . . .	You should . . .
information about a bill	pull the ledger and answer the questions yourself.
information about fees	follow office policy. In most veterinary offices, you can quote set fees, such as those for distemper and other vaccinations. Usually, you can give a range for fees that vary slightly. "Mrs. Smith, the charge for Buster's neutering will probably be between $30 and $50." Sometimes, you might have to explain that the veterinarian must examine the animal before setting a fee for other services.
to report on satisfactory progress or treatment	take a message and say "I will make sure Dr. ___ gets this information."
to report on unsatisfactory progress or treatment	take a message and let the client know that either you or the veterinarian will get back to her.
to request test results	with the veterinarian's permission, give favorable results. The veterinarian should call to report unfavorable results.
to complain about care or fees	explain any points you can, or offer to call back after you have checked the chart. If you cannot satisfy the client easily, it is best to have the veterinarian call the client back.

As you can see, coordinating calls is another area where you have to use good interpersonal communication skills.

Outgoing Calls

Answering the phone is only one side of the coin. As part of your veterinary assisting job, you will also have to make calls. This task requires all the courtesy and tact of answering the phone, plus a few other traits and skills.

First, you should be organized. Be especially careful to organize your call when you have more than one item to discuss. It is helpful to jot down a few notes to prompt your memory.

You should think of time—not only the time of day but also the time you have available to speak without interruptions. The time of day is important to a client whose ringing phone interrupts a sound sleep. And you can hardly call to order supplies when the reception room is full and the phone is ringing.

Frequently, you will make outgoing calls to clients and other veterinarians your veterinarian wants to speak to. In these cases, your role is straightforward. You will say something like, "This is Liz from Dr. Doolittle's office, returning Mr. Green's call." If the person is not expecting the call, you might say something like "Dr. Doolittle would like to speak to you about Togo's diet." Then you will ask the person to hold for the doctor.

You might also make calls to:

- remind clients when boosters and vaccinations are due
- confirm appointments made far in advance
- advise clients that the veterinarian is running behind schedule, and ask them to delay coming in for a stated amount of time

Sometimes, you will be asked to relay messages for the veterinarian. For example, you might call to reassure a client that a pet is recovering well after surgery or an illness. Or you might have the unpleasant assignment of calling clients to remind them to pay past due accounts.

Collection Calls

Most animal hospitals now expect payment when services are rendered, but sometimes bills do accumulate. Some management consultants strongly advise against telephone collection calls because they may interrupt the client at an inopportune time or embarrass the client in front of others present. Sometimes collection calls are unavoidable, though. Some clients may feel that promising to pay is all that is necessary. Unless you follow up, many clients still will not put the check in the mail.

Just be careful how you go about a collection call. And *never* take it on yourself to handle past due accounts without your employer's approval. Always advise your employer of any problems that develop. Never make a collection call within earshot of others, especially clients in the reception area. Not only is this in bad taste, but it could lead to a lawsuit against your employer. Also, be sure not to violate criminal statutes or telephone company rules. Violations include the following practices:

- calls placed at odd hours
- repeated calls (considered harassment)
- calls to debtors' friends, relatives, employers, or children
- calls making threats
- calls falsely asserting that credit ratings will be hurt
- calls demanding payment for amounts not owed
- calls that frighten, abuse, torment, or harass another person

People frequently react in a hostile manner when reminded about a past due account, so try to approach the subject with sensitivity. Always speak directly to the person responsible for the bill.

If you must leave a message, do not mention the nature of the call. Just leave your name and number, and call back another time if the person fails to return your call.

Trying to collect past due accounts is not a pleasant task, but you can make it easier by beginning the conversation with polite "small talk." Asking questions such as "How are you doing?" and "How did you like the big blizzard last week?" will put the person in a more receptive frame of mind.

Ask courteously for payment. You must be firm, but not too blunt. You can usually avoid offending the person by assuming that the nonpayment is due to an oversight. Your manner should be pleasant and reflect your confidence that the client is not deliberately avoiding his or her obligation. You can comment on how good the person's record has been in the past, if appropriate.

> VETERINARY ASSISTANT: Your payment record with us has always been fine until now. I was concerned, not having heard from you. I wanted to discuss the status of your account.

This will certainly get some response. Hear the client out; do not interrupt the client during the call. If the client pleads hardship or launches into complaints, listen carefully and promise her that you will discuss the matter with your employer. Never show irritation in your voice or appear to be scolding the client. You simply want to find out why she has not paid or answered the collection letters you have previously sent.

The person will be more apt to cooperate if you indicate that the office can be flexible in a payment plan. For example, you can suggest a partial payment now, followed by small monthly payments until the balance is paid. If the client promises to pay, ask when you can expect the check. Have the client write down the payment agreement. For instance, if the client is to send in half on Friday, ask her to write this information down, along with the veterinarian's name and address and dates of additional payments. Then ask the client to read back to you what she has written to make sure there is no misunderstanding. It is a good idea to follow up on the phone conversation with a brief letter (Figure 7-5).

Documentation is very important with collection calls. Make sure you keep note of when "the check is in the mail"—or when it is supposed to be—so that you can keep an eye on problem accounts. Most people are as good as their word, but occasionally follow-up calls may be necessary to "refresh" a person's memory about payment agreements. Use a **tickler file** to follow up on the commitment dates. (A tickler file is like an appointment book, only it has slots for mail and telephone messages; they are organized by days of the month. Reminders are placed in the file on the day of the month concerned. Your computer might also have an automated tickler program. Tickler files should be checked daily.)

Remember that the goal of a collection call is to try to get payment without losing or offending the client. An uncollected bill is insignificant compared to the damage that can result from a dissatisfied client's criticism. Word of mouth can spread criticism faster than praise!

Shelly Thompson, D.V.M.
303 Main Avenue
Any City
(000) 555-0000

May 12, 20—

Mr. Stine G. Client
Avenue of Accounts Uncollected
Anycity

Dear Mr. Client:

Thank you for spending a few minutes of your busy day to speak with me about your account. I am glad we were able to work out an agreeable payment plan.

I will look for your first installment of $20.00 in the mail early next week. After that, you can send a payment of $10.00 on the first of every month for 10 months (June through March). At that point, your current balance of $120.00 will be paid in full.

Please contact me if you have any questions.

Sincerely,

Liz Billings

Liz Billings

Veterinary Assistant

FIGURE 7-5   Collection calls are usually more successful when followed up by a letter.

Off-Hours Phone Coverage

Someone must be available at all times to cover (answer) the phone. Veterinarians usually subscribe to professional answering services to provide coverage during off hours such as nights, weekends, holidays, and vacations. The service screens client calls according to the veterinarian's orders. Instead of a service, some veterinarians use their own answering machines, electronic answering devices, or voice mail systems in which clients can leave messages for the doctor or staff to handle directly. You might even be asked to record the veterinarian's message on the answering machine.

> *Sample message:* Hello. You have reached the Animal Care Hospital. We are sorry we cannot take your call right now. If you have a medical emergency, please call Dr. Morgan at 555-4321. If you would like to speak with Dr. Thompson or make an appointment, please call back tomorrow between the hours of 8:30 and 5 o'clock.

MAKING APPOINTMENTS

Scheduling Appointments

One of the most important administrative duties of a veterinary assistant is scheduling appointments. You will be responsible for pacing the veterinarian's day according to his or her preferences. Schedule too many patients, and the result is chaos. Schedule too few, and the office loses profits. Your goal: to schedule the day just right, so that no one's time is wasted, the clients are happy, the patients are well cared for, and the office prospers.

Types of Scheduling

There are three general types of appointment scheduling:

1. *Wave scheduling.* With **wave scheduling** the total number of patients to be seen in one segment is scheduled at the same time. For example, if the average time for each patient is 15 minutes, four patients are scheduled for each hour and are seen in the order of arrival. You might schedule four patients for the hour from 10 A.M. to 11 A.M. This method helps to adjust for any deviations from the schedule such as no-shows, late arrivals, or walk-ins. Although some patients may be seen later than scheduled, each hour is usually started and finished on time.

2. *Flow scheduling.* Using **flow scheduling,** patients are scheduled for 15-minute intervals. Provisions are made for longer appointments. This method usually results in the best control of scheduling and in a shorter waiting time for clients and patients. However, once you are behind schedule, it is almost impossible to catch up.

3. *Fixed office hours.* Clients come to announced **fixed office hours** with their pets whenever they wish. On arrival, they register or sign in and their pets are seen on a first-come, first-served basis. Obviously, with this system, it is difficult to control the flow of patients and make efficient use of resources and staff.

Whatever the type of scheduling, you always need to consider the time necessary for the examination or procedure and the availability of examination or treatment rooms.

REMINDERS

- Many clients like to receive a reminder when it is time for a vaccination booster. Reminders are usually sent about a month before a pet's vaccinations are due.

- Some offices use telephone reminders. A call saves postage; and even better, it guarantees that the client gets the message.

- Whether you remind by mail or phone, always check the client's file first to make sure the pet has not been put to sleep.

CITY ANIMAL HOSPITAL
1234 ANY ROAD
ANYCITY, STATE 99999
PHONE: (101) 555-1234

ANYCITY, STATE
PM
06 MAR
20XX

A reminder that BUSTER is due for:

ANNUAL WELLNESS EXAM W/ VACC.	MAR 20XX
PARVOVIRUS VACCINATION	MAR 20XX
BORDETELLA BOOSTER	MAR 20XX
HEARTWORM EXAMINATION	MAR 20XX
DHLPC BOOSTER	MAR 20XX
INTERSINAL PARASITE FECAL EXAM	MAR 20XX

Hospital Hours: FOR APPT. CALL (101) 555-1234
    Monday through Thursday 7:00 AM to 8:00 PM
    Friday 7:00 AM to 8:00 PM
BOARDING AND GROOMING ONLY CALL (101) 555-5678

MR. AND MRS. DOGOWNER
123 LEFT LANE
ANYCITY, STATE 99999

		DAY DATE	MON 6/23
Julie Simon, 555-0001; Persian; Amos; Eye Infection	**8:**		00
			(15)
			30
			45

FIGURE 7-6   An example of an appointment book entry for a scheduled appointment.

The Appointment Book

Many styles of appointment books can be used in animal hospitals. If you are given your choice, be sure to pick one that allows enough space for the number of appointments your office permits in a particular time span (Figure 7-6). And make sure that each appointment slot has space for the following data:

- client's name and phone number
- pet's breed and name
- a brief notation of the reason for the visit

When you have your appointment book, go through it immediately and block off the times when the veterinarian will not be available for appointments (Figure 7-7). Use red pencil so that the marking stands out. You do not want to accidentally schedule an appointment during the doctor's vacation! Unavailable times might include:

- holidays
- vacations
- personal appointments
- Sundays
- lunch breaks
- surgery
- days off
- evenings

This advanced preparation of the appointment book is sometimes called **establishing the matrix**.

The following suggestions should help you keep an accurate, effective appointment book:

		MONDAY, MAY 9					TUESDAY, MAY 10		
8	00				8	00			
	15		*Out*			15		*Out*	
	30					30			
	45		*Accountant*			45			
9	00				9	00			
	15					15		*Dentist*	
	30					30			
	45					45		*Appt.*	
10	00				10	00			
	15					15			
	30					30			
	45					45			
11	00				11	00			
	15					15			
	30					30			
	45					45			
12	00				12	00			
	15		*Lunch*			15		*Lunch*	
	30					30			
	45					45			

FIGURE 7-7   Remember to block off the times when the veterinarian will not be available for appointments.

- Read back appointment data taken over the phone so that there is no mix-up. "We will see you at 9:30 A.M., Monday, May 9th."
- Confirm appointments the day before if they are longer than 15 minutes. This helps avoid time (and revenue) lost due to missed appointments.
- Ask clients to call if they cannot keep an appointment.
- Give new clients clear directions to the office so that they do not get lost and arrive late.
- Discourage walk-ins except for emergencies, such as injury or acute illness.
- Give appointment cards whenever possible (Figure 7-8).
- Keep the doctor aware of waiting clients and patients, especially when she is running behind schedule.
- Make sure you write every appointment down in the book, and always indicate when an appointment is canceled.
- Record no-shows and cancellations on the patients' charts and call or send notices to these clients asking them to reschedule.
- For cancellations, neatly cross out the original appointment and give the time to another patient, writing the new appointment above the old. (Because the appointment book can serve as a legal document if a client should sue the veterinarian, erasing or whiting-out the canceled appointment could create legal complications.)

ANIMAL CARE HOSPITAL

Fido

_____

HAS AN APPOINTMENT ON

Tuesday   *June 27*
_____

DAY   MONTH   DATE

AT _____ A.M. *4:30* P.M.

Please telephone one day in advance if
you are unable to keep the appointment.
Telephone 555-0000.

FIGURE 7-8   When you give out appointment cards, clients are more likely to remember their appointments.

Organizing Appointments

If your appointment book has slots for three patients every 15 minutes, should you schedule three patients every 15 minutes? Not necessarily. Appointments are scheduled according to the type of veterinary treatment, availability of examination and treatment rooms, and available time. You should schedule each patient for the amount of time needed for that particular appointment. That time can range from just a few minutes for a vaccination to several hours for surgery. Some lengthy procedures have periods in which the veterinarian's attention is not necessary. In those cases, you can schedule short appointments within the time frame of the longer one. You might also want to block off a periodic free period of 15 minutes to allow for any spillovers in the schedule or emergencies. Sometimes clients request a specific veterinarian, and sometimes an appointment can be made for a veterinary technician, for suture removal, for example.

To make the best possible use of the veterinarian's time, you must know how long each procedure usually takes and then consider that time as you schedule the appointment. If you schedule carelessly—making appointments for patients without considering what is to be done—you will probably end up with a full waiting room. This leads to dissatisfied clients and an unhappy employer.

To make scheduling easier, you should chart the time it takes to perform various procedures in your office. That way, you can tell at a glance if you can squeeze in Jimmy Cardin's cat's allergy shot in the afternoon, or if you will have to cancel Louisa Oswald's parrot's checkup so that the doctor can set Jenny Smith's dog's broken leg this morning. Veterinarians all have different paces, so you will have to consult your employer to determine specific times.

Scheduling House Calls

Clients with small animals usually bring them into the office for treatment. You should encourage this, because the veterinarian can better attend to animals in an office with the necessary equipment and testing facilities. Large animals, however, are often treated on their own turf. A sick cow, hog, or horse is difficult to transport! So you need to be familiar with procedures for scheduling house calls.

Your employer will have a policy you should follow in regard to home visits. You will probably be asked to take a detailed message so that the veterinarian can decide if and when to make the house call. Write down this information:

- full name of client
- name and breed of animal
- address, with clear directions for getting there
- phone number
- reason for the call

Some offices use a standard message form for house calls (Figure 7-9). As soon as you know the time the veterinarian will visit, call to let the client know.

FIGURE 7-9 Make sure your house call message is clear and complete so that your veterinarian can make the correct decision.

To: *Dr. Doolittle*		
Date: *5/12*	Time: *3:45*	A.M. ☒ P.M.

WHILE YOU WERE OUT

Miss Jessica McMahon

of: *Box 487, Rt. 23, Thistown*

Phone: *555-3333*

Telephoned	☐	Please Call	☐
Came to See You	☐	Will Call Again	☐
Wants to See You	☒	Rush	☒
Ret'd Your Call	☐	Called Again	☐

Message *4 Dairy cows sick w/diarrhea; no appetite; gas. Please come A.S.A.P. First farm on right after Thistown corners*

Signed: *Liz*

Try to cluster these in the same area to save the veterinarian the time and trouble of driving back and forth across town.

GREETING CLIENTS

Now that you know how to make appointments, you are ready to learn about another important aspect of the job: reception. If all you had to do was sit behind the desk and say hello to clients, reception would be easy. But chances are that you will have to perform many other duties at the same time you greet clients. And that is not easy—though the challenge can be very rewarding.

Any number of events might change the pace of the day you so carefully scheduled. A seemingly routine ailment might turn out to be complicated and time-consuming. Another veterinarian might drop by for an emergency consultation. Or your veterinarian might get held up in traffic. No matter what the cause, tragic or not, you must be flexible enough to accommodate the unexpected. Throughout the sometimes chaotic changes in the appointment schedule, you must remain calm and cheerful to greet clients and patients and soothe those in distress. In short, you are responsible for creating and maintaining a good social, physical, and professional atmosphere in the veterinary office.

First Impressions

Would you buy groceries at a store with old, torn advertisements in dirty windows if you could go to a store with bright, new ads on sparkling clean windows? If you are like most people, you will assume that the old ads and dirty windows mean the groceries will be old and dirty, too. Based on your first impression of the outside, you will pick the clean, neat store. The same applies to veterinary offices (Figure 7-10). A carelessly kept waiting room suggests to clients that the veterinarian is careless, too. And who wants to trust their beloved pet to a careless veterinarian? If you do not keep the office as clean, neat, and odor-free as possible, your employer will lose business.

Your reception duty starts before the first patient of the day arrives and continues until the last one leaves. Make sure you arrive early enough to make certain the reception area is clean, neat, and odor-free, in addition to your other early morning duties. Make sure magazines are tidy and coffee cups and other refuse are discarded. Sweep, dust, or pick up as needed. Tables and other furniture should be wiped down each day with disinfectant to protect clients and patients from any contagious diseases (viruses, for example). Many veterinary assistants check the waiting area after every few patients. If this seems extreme, remember that every client—even the last one of the day—expects the office reception area to be clean and inviting. And, as you know, even the most well-trained pets might have accidents if they are scared and nervous. Even if you are not the one responsible for cleanup, you are respon-

FIGURE 7-10   The veterinary office should make a good first impression on clients. *(Setting courtesy of Airpark Animal Hospital, Westminster, MD)*

sible for calling the appropriate staff member to take care of messes. (If you work in a small office, cleanup will probably be your responsibility.)

To control odors, you should begin each day with a thorough airing of the office. If there is no air conditioning available, then ceiling fans, opened *screened* windows (you do not want to offer any opportunities for wild animals to sneak in or for pets to sneak out), and plenty of space will help ventilate the area while it is in use. You can use deodorizers to help keep the air fresh smelling, but they should never replace a thorough airing. Heat tends to magnify odors and make animals uncomfortable, so make sure the reception area does not get too hot. Set the waiting room thermostat no higher than 68 degrees Fahrenheit or 20 degrees Celsius.

First impressions are not limited to the office itself; your appearance is important too. We have already discussed the importance of your pleasing telephone personality in creating a favorable first impression. When you greet clients, you should be just as pleasant and charming in person. Of course, no veterinary assistant *tries* to irritate or offend clients, yet it happens. How do you ensure that you will make a favorable first impression when there is no way to tell what offends or irritates someone you do not know? Actually, there is no way to please everyone all the time, so you should not expect to. But you can do your job best by focusing on two areas: professional appearance and good interpersonal behavior.

Professional Appearance

After the initial phone contact, when a client first sees you he or she will notice how you look. (We have discussed this issue before, but it is so important that it should be repeated.) Most veterinarians ask their staff to wear some sort of uniform. It might be a smock worn over your street clothes, a combination of top and slacks, or some sort of medical uniform. Whatever your clothing requirements may be, always make sure that your clothes are clean, neat, and in good condition. If you do not wear any sort of uniform, dress with good, professional taste.

Suppose you were bringing a family member to a physician's office. How would you expect the staff to be dressed? Would you feel confident if you were greeted by a woman in a skimpy top and miniskirt? Would you be comfortable being treated by a doctor wearing jeans and a T-shirt? While these outfits might be fashionable at certain times and in certain social circles, they are never appropriate in a professional medical setting, including veterinary practices and animal hospitals. If your office does not have a uniform or a dress code, use common sense when choosing an outfit. For men, slacks and a button-down shirt are recommended. While a necktie expresses a more professional air, it can get in the way if the wearer is in regular contact with animals. For women, slacks or a modest skirt and a blouse are appropriate. While jewelry is certainly attractive, loose fitting necklaces or bracelets can pose a danger when handling animals, and shiny dangling earrings might also prove too tempting a target for certain animals. Stud earrings are preferred. Most necklaces and bracelets should be avoided.

If you are unsure of what to wear, pay attention to what the rest of the staff is wearing on your first visit, and ask your supervisor what is appropriate in your setting.

> *Replace scuffed briefcases, worn or dirty purses, rundown heels. Leave your gum at home."*
> —*From* The Job Survival Instruction Book

Your personal hygiene will be noticed, too—neatness and cleanliness are essential. Keep hair and nails tidy, and jewelry and makeup simple. Keep perfumes and colognes to a minimum. (Some clients find them offensive; others may be allergic to heavy scents.) Smoking offends many people, and since it can be a health hazard, most veterinary offices do not allow it. But if yours does, avoid smoking in front of clients. Avoid body piercings, other than pierced ears (this may create a less-than-professional image of the practice in the mind of the client).

Think of every detail of your appearance and consider what kind of impression you make. Something as seemingly insignificant as your shoes could send an unfavorable message. For example, spike heels might suggest to a client that a woman is impractical and inefficient. A shirt unbuttoned to the waist might suggest that a man is unprofessional. Perhaps it is not fair, but if you want to succeed as a veterinary assistant, you should consider what people think. *The care you take in presenting a clean, neat, efficient appearance will go a long way with clients who feel nervous about their pet's care.*

Interpersonal Behavior

You might sound personable on the phone, have a pleasant appearance, and still make a poor first impression on a client. All you have to do is ignore someone who has just arrived. Put yourself in the client's situation and imagine how annoying it is to arrive on time, only to be completely ignored by the veterinary assistant. As we have said before, empathy—the capacity to understand others' feelings—is an essential character trait for the veterinary assistant.

The very *first* thing you must do when a client arrives is to acknowledge her. Even if you are on the phone or busy with something, acknowledge the client with a wave or a smile. If your duties require you to leave the reception area, you could leave a sign-in sheet handy so that clients can register their pets, and/or a bell so that they can get your attention. Though not as effective as an immediate personal greeting, the register or bell will let clients know you have not forgotten them. As soon as you can, offer a personal greeting.

> VETERINARY ASSISTANT: Hello, Mr. Jones. I am sorry to keep you waiting. Would you like to sit down? Dr. Thompson will be with you in just a few minutes. Meanwhile, feel free to help yourself to some coffee.

For clients who have not been to the office in some time, verify their addresses, telephone numbers, and any other necessary information as soon as possible.

> VETERINARY ASSISTANT: Mr. Jones, are you still at the same address and phone number in Dalton? Is Smudge up-to-date on her vaccinations?

For new clients, your greeting can be just as friendly. Begin by locating both the client's and the patient's name in the appointment book.

> VETERINARY ASSISTANT: Are you Mrs. Simpson? It is nice to meet you. I am Liz. Will you please have a seat? I will be right out to ask you some questions about Rocky.

Of course, you do not have to use the same words each time. The important thing is to be natural, friendly, polite, and *genuine*. Reception is an opportunity to establish a positive relationship with both clients and their pets. Being enthusiastic about your work will help make the veterinary office a more relaxed setting and more enjoyable for clients, patients, and other employees.

Monitoring Patients

The veterinary assistant should **monitor** patients—that is, keep track of who is there for what. Although it is usually best to take prompt clients and patients in the order of their appointment times (or, in the case of patients who are a little late, their order of arrival), there are exceptions. Clients who are so late that the doctor really does not have time for them can present a sticky situation. You do not want them to get angry, but you do not want to inconvenience other clients, either. If you really cannot fit the patient in, use tact in offering another time.

> VETERINARY ASSISTANT: I am sorry that your car broke down, Ms. Bensalem, especially since Dr. Thompson is scheduled to perform surgery this afternoon and still has several other patients to see first. Could you possibly bring Felix in at nine tomorrow morning?

Clients whose pets are very ill or contagious should be escorted to an examination room as soon as possible.

Handling Delays

Another important factor in a client's first impression of a veterinary office is the length of the wait. Waiting with animals can be uncomfortable and can make clients nervous. When Mrs. Smith makes an appointment for Rex for 2:00 P.M., she expects to see the veterinarian at 2:00 P.M. So, an important part of veterinary assistant–client relations is honesty about delays. Most people understand a 5- or 10-minute delay and will wait patiently. Any longer than that, and they watch the clock with increasing annoyance. You can avoid potential dissatisfaction by letting clients know when the wait will be considerable. If the veterinarian is going to be delayed, explain this to the clients and give them the option of canceling, waiting, or rescheduling. Clients who will have the longest wait should be informed of the delay first. And, as you learned earlier, clients with later appointments appreciate your calling to inform them about delays of a half an hour or more.

Never be vague or noncommittal about how long a wait will be. At least give a reasonable estimate.

> VETERINARY ASSISTANT: I am sorry, Mr. Jones, but I am not sure exactly how long Dr. Thompson will be in surgery. It usually takes an hour, and she went in 40 minutes ago.

To show your concern, you might suggest the client take his pet for a stroll around the back of the building, or turn on the television set (if available). Clients appreciate a personal touch, which can make them feel more important as individuals.

In addition to helping clients, your personal touch will make your own work easier, more pleasant, and more rewarding. For instance, if the veterinary assistant notices that the office is always behind schedule, the office staff should analyze the situation. Is enough time being scheduled to complete each procedure? Does the veterinarian have trouble staying on schedule? Once the problem is identified, the staff can take steps to solve it.

EUTHANASIA

Euthanasia is the medical term for ending an animal's life humanely and painlessly when it is suffering from a painful, incurable disease or condition. Although various methods of euthanasia are used in different settings, the most common form for companion animals in a practice or hospital setting is a chemical solution injected with a hypodermic needle. This will cause the animal to go into an almost immediate deep and irreversible state of unconsciousness, and then death.

The veterinarian may offer euthanasia as an option in cases of extreme illness or incapacitation, but the decision to euthanize can be made only by the client.

Common Client Concerns

It is not unusual for clients to be hesitant when it comes to euthanasia, even when they know it is the best possible option for the animal. To make the situation easier, it is important to understand what kinds of concerns they might have, and then address them as necessary.

Using the Pet for Experimentation or Selling the Body.   Urban legends, movies, and other stories might lead some clients to believe that the practice or hospital will seek to profit from the death of the animal by using the carcass for either reputable or questionable scientific purposes. Any clients expressing reservations about euthanasia for this reason should be assured that their pets will be treated with the same respect in death as in life, and that their bodies will not be sold or used in any way after being put down, but rather disposed of respectfully.

Being Present during Euthanasia.   Generally speaking, clients may choose whether not to be present during euthanasia. Some find it comforting to be there, both for themselves and the pet. Others feel the moment would be too much for them to bear. Either way, the choice of whether to be with the animal should be made strictly by the client.

Where the Euthanasia Will Take Place.   In the case of companion animals, the process typically takes place at the practice or hospital. Depending on the policies where you work, at-home euthanasia may also be available. Be sure to speak with the veterinarian and your supervisor to find out if this service is offered, and what special procedures, preparations, and policies are involved. Livestock, zoo animals, and other types of large animals are generally euthanized at their own location.

Disposition of the Remains.   It is standard practice for the hospital or office in which the euthanasia takes place to dispose of the animal's body after death. Clients generally have the option to take the deceased pets with them, for burial in a pet cemetery, cremation, or burial at home (when it is allowed by law).

Grief

It is very common for clients to experience pain, sadness, and grief at the loss of a pet. In many situations, the animal was regarded as a member of the family, and its loss is felt as deeply as the loss of a child or sibling. Being in such a powerful state of sorrow can also bring back other sad memories, making a person feel even more depressed. As a member of the veterinary team, you might feel a certain degree of grief yourself over the death of a patient, especially one you have been seeing for years, or who you treated in a hospital setting.

A person who loses a pet, whether through euthanasia, illness, accident, or any other method, may experience any or all of the following:

- *Shock*. In many cases, the loss of a pet does not result from a prolonged illness, but rather a severe injury, short-term illness, accident, or other form of sudden death. In these situations, the clients rarely have any time to "prepare" themselves for the loss of the pet, and may find themselves unable to accept the animal's death right away.
- *Denial*. For some people, the sense of shock can lead to denial, not truly believing that the animal is gone. Many expect to see or hear the animal when they get home, or in some cases, that the animal is somehow alive somewhere. If feelings of denial do not pass, the person should seek professional counseling.
- *Anger*. It is a natural (although not always justified) reaction for many people to become angry in the midst of a traumatic event. A client might be angry at the veterinarian and veterinary staff for not being able to save the pet, the person who hit the animal in the event of a car accident, God, society, or any of a number of other people, or institutions.
- *Sorrow*. Most, if not all, of the clients you deal with will feel sad over the loss of their pets. This reaction is natural and normal, and is often accompanied by tears. While your natural reaction might be to comfort the person, be careful about touching clients. It is better to offer soothing words or condolences, and to listen if there is something the client wants to say. Be sure to keep a box of tissues handy for grieving clients.
- *Depression*. While it is normal to feel sad after the loss of a pet, some people enter such a strong state of grief that they fall into a depression. Anyone still feeling an overwhelming sense of sadness and loss over a period of time should speak with a professional counselor.
- *Guilt*. Sometimes people feel guilty, thinking that there might have been something they could have done to prevent the pet's death, or that the choice to euthanize was incorrect. Left unresolved, feelings of guilt can prolong the recovery process.

You should be prepared to deal with a client expressing any or several of these emotions before, during, and after the euthanasia process. Try not to take any client comments or outbursts too personally, as she may not be in an entirely rational state of mind. Be courteous, kind, and respectful, and have any pet loss information, including support group and grief counselor listings, available for clients to take with them.

Dealing with Grief.   As with any other emotionally traumatic event, it takes time to get over the loss of a beloved animal. How long the recovery process takes differs from person to person. Helpful strategies that aid recovery include:

- Talking about the loss with a close friend or other family members.
- Talking with other people who have lost pets. Support groups exist in many areas, and your practice or hospital should have information about local resources. If not, do a little research and compile a list of local support groups, grief counselors, and on line resources.
- Allowing themselves time to recover.
- Taking proper care of themselves, including eating right, and getting adequate amounts of exercise and rest. Getting ill can cause the grief period to last longer.
- Remembering the pet and what it meant to their lives.
- Getting another pet when the person feels ready, as a new pet can help a person to recover from the loss of a previous pet. The question of "when" varies from person to person, and in many cases, the answer is "never."

Other resources, such as pet loss hotlines and professional grief counselors are available in many areas. There are also a number of books about coping with the loss of a pet, as well as on-line support groups. Check with veterinarians, the local humane society, and animal hospitals to find out what resources are available in your area. The on-line resources at the end of this chapter may also prove helpful to you in your search.

SUMMARY

When you answer the telephone at the veterinary office, you become the voice of the practice to the person on the other end, so you need to exhibit the respect, compassion, and professionalism that you would expect to hear if you were the person making the call. You should be familiar with the office's method of screening calls, making appointments, and taking messages. When it comes to calling others, be sure you understand why you are calling, have all the information you might need to answer questions, and always be respectful, whether you are reminding a client about an appointment, asking a question, or requesting payment on a bill.

When making appointments, you should know whether your office uses wave scheduling, flow scheduling, or fixed office hours, and utilize the appointment book to ensure that there is a record of exactly who is coming in and when. Be aware of how long each appointment will last, to avoid long waits for other patients. When there are delays, be sure the veterinarian knows patients are waiting, and let the patients know about how long a wait they can expect.

The reception area should be kept neat and clean, and you should always maintain a professional appearance.

Euthanasia is a common medical procedure that requires the veterinary assistant to be simultaneously professional, tactful, and compassionate. Special arrangements may need to be made when a patient is to be euthanized. It is not uncommon for clients to have a strong emotional reaction to the death of a pet. The veterinary assistant should be aware of the range of emotions that the client might experience, and have pet loss information materials available for clients who want it.

CASE STUDY

Stephanie, a veterinary assistant at Dr. Brady's office, answered the phone one morning. "Good morning, Dr. Brady's office. Stephanie speaking, can I help you?"

"Hello, this is Mr. Munson. I'd like to make an appointment for my iguana, Stilson."

"Okay, is Stilson feeling all right?"

"Oh, he's fine, it is just time for his annual checkup."

"That is good to hear!" Stephanie said, checking the appointment book for availabilities. "Do you prefer mornings or afternoons?"

"Afternoons are much better for me, thank you."

"Can you come in Thursday at 2:30?"

"That would be fine."

"Okay, I've got you down for Thursday," Stephanie said, filling in his name in the appointment book. "Oh, our records show that you have an outstanding invoice from Stilson's emergency visit just over two months ago. Did you receive a copy of the bill?"

"Oh my, I forgot all about that! Can I bring a check with me when I come in on Thursday?

"That would be fine, thank you."

"Okay, goodbye."

"Goodbye, Mr. Munson."

- How did Stephanie's handling of the conversation help her to get the specific information she was looking for as quickly and completely as possible?
- Was it appropriate for her to bring up the outstanding bill while making the appointment?
- Which form of scheduling does this office most likely use?

REVIEW

1. Based on the following conversation that occurred at 11:45 A.M., fill out a message form like the one that follows the conversation. You may use today's date.

VETERINARY ASSISTANT: Dr. Morgan's office. Anne speaking. Can I help you?

CALLER: This is Maria Jones. I would like to speak with Dr. Morgan about my cat.

VETERINARY ASSISTANT: Oh, yes, Mrs. Jones. How is Tabby doing after her surgery?

CALLER: She has had a bit of bleeding around her stitches. I would like to talk to the doctor about it.

VETERINARY ASSISTANT: I am afraid Dr. Morgan is unavailable right now, but I would be glad to leave a message for him to call you. Could I have your phone number, Mrs. Jones?

CALLER: Yes, it is 555-1234. Please tell him Tabby bleeds whenever she moves.

VETERINARY ASSISTANT: And when did the bleeding start, Mrs. Jones?

CALLER: Eight this morning. When can I expect the doctor to call?

VETERINARY ASSISTANT: He will call you as soon as possible. Please try not to worry, Mrs. Jones.

CALLER: Thank you, I'll try! Goodbye.

VETERINARY ASSISTANT: Thank you for calling, Mrs. Jones. Goodbye.

To:		
Date: Time:		A.M. P.M.
WHILE YOU WERE OUT		
Phone:		
Telephoned	☐ Please Call	☐
Came to See You	☐ Will Call Again	☐
Wants to See You	☐ Rush	☐
Ret'd Your Call	☐ Called Again	☐
Message		
Signed		

2. The veterinary assistant in the following conversation makes mistakes in every response. Fill in what you think should be said. Make up any details you think are needed.

VETERINARY ASSISTANT: Hello, please hold.

a. _____

_____

VETERINARY ASSISTANT: (After three minutes) Yes, can I help you?

b. _____

_____

CALLER: I'd like to make an appointment.

VETERINARY ASSISTANT: Just a minute. (Puts caller on hold for two minutes.)

c. _____

_____

VETERINARY ASSISTANT: Okay. What did you say your name was?

d. _____

_____

CALLER: John Doe.

VETERINARY ASSISTANT: How is next Friday at three?

e. _____

_____

CALLER: Actually, my cat got into a fight with another cat last night. He's cut up pretty bad. I really think the vet should see him today.

VETERINARY ASSISTANT: Oh, an alley cat on the prowl? Well, bring him in anytime, then, Mr. Doe. What did you say his name was?

f. _____

_____

CALLER: Tiger.

VETERINARY ASSISTANT: Well, that figures! See you later. (Hangs up, laughing.)

g. _____

_____

3. The veterinary assistant makes mistakes in each of the following situations. Write what you think should be said or done.

a. An incontinent elderly dog just had an accident on the waiting room floor. The veterinary assistant says to the client, "I am sorry, but could you please clean up the mess? The boy who usually does it is on break."

_____

_____

b. The veterinary assistant is on the phone and pays no attention to the line of clients and pets waiting to check in.

_____

_____

c. The veterinarian has been unavoidably delayed for one hour, but the veterinary assistant says nothing to the waiting clients.

_____

_____

d. A client comes into the office to complain about a bill and yells at the veterinary assistant. The veterinary assistant gets angry and says, "If you cannot afford to keep your pets healthy, you should not be allowed to have them."

_____

_____

4. Create a log like the one that follows and enter the listed appointments from Dr. Doolittle's schedule.

Office opens 8:00 A.M.

Surgical procedures in the early morning

Lunch from 1:00 P.M. to 2:00 P.M.

A college course from 2:00 P.M. to 3:00 P.M.

Rounds to check on hospitalized animals from 4:00 P.M. to 5 P.M.

8	00	
	15	
	30	
	45	
9	00	
	15	
	30	
	45	
10	00	
	15	
	30	
	45	
11	00	
	15	
	30	
	45	
12	00	
	15	
	30	
	45	
1	00	
	15	
	30	
	45	
2	00	
	15	
	30	
	45	
3	00	
	15	
	30	
	45	
4	00	
	15	
	30	
	45	
5	00	
	15	
	30	
	45	

Indicate whether the following statements are true or false.

5. When a patient is euthanized, the client is required to take the remains.

6. Only medical personnel may be present during euthanasia.

7. Shock is a normal reaction to the loss of a pet from an accident.

8. A prolonged state of sorrow and grief might indicate a state of depression.

9. For most people, talking about the loss of a pet only makes grief worse.

10. Proper nutrition and exercise are important to the grief recovery process.

ON-LINE RESOURCES

Mastering Telephone Etiquette (from Dummies.com)

This article from Dummies.com offers tips for proper phone usage when making and receiving calls, and when dealing with angry callers.

<http://www.dummies.com/>
Search Term: Telephone Etiquette

Using the Telephone (from Careers-Portal)

This article explains the advantages of using the telephone at work, what skills a person should have when making calls, and what can go wrong.

<http:// www.careers-portal.co.uk/>
Search Term: Telephone

How to Prepare for a Veterinary Office Visit (from About.com)

Although this article is directed toward pet owners, a veterinary assistant can make use of the tips and advice by suggesting them to clients, ensuring that they arrive at the office completely prepared for the visit.

<http://vetmedicine.about.com/>
Search Term: Veterinary Office Visit

Pet Loss Support Pages

<http://www.lightning-strike.com/>
<http://petloss.com/>
<http://www.pet-loss.net/>

CHAPTER 8

Ethics

OBJECTIVES

When you complete this chapter, you should be able to:

- identify ethical rules of conduct for veterinarians and members of the veterinary staff
- explain what behaviors and practices are not allowed in the practice of veterinary medicine
- explain when the veterinarian-client-patient relationship begins, the responsibilities of those involved in it, and how it can be dissolved

KEY TERMS

ethics	confidentiality	veterinarian-client-patient relationship (VCPR)
jurisdiction	substance abuse	
slander		

INTRODUCTION

Veterinary staff members, like veterinarians and any other medical care professionals, must adhere to an ethical standard of behavior. This chapter presents a general overview of **ethics**—the rules of moral conduct all members of the veterinary staff should follow.

PUTTING THE NEEDS OF THE PATIENT FIRST

The veterinarian's job, first and foremost, is to relieve an animal's suffering, illness, or disability. The veterinary staff's primary obligation is to aid the veterinarian to fulfill this duty.

Right to Choose Clients and Patients

A veterinarian may choose to accept or decline any client or patient at any time, as long as the refusal of such service does not pose an immediate threat to the patient's life and does not result in prolonging the animal's suffering. However, once a client is taken on, the veterinarian, and by extension, the veterinary staff, must provide service. It is never acceptable to neglect your patients.

Emergency Care

Veterinarians, and only veterinarians, are allowed and required to provide essential services when an animal's life is at stake. If you are a qualified veterinarian, you have a moral duty to provide care to save an animal's life or to relieve its suffering. By the same token, however, only a veterinarian should provide such service. Nonveterinary members of the veterinary staff are not qualified to provide medical care in an emergency situation any more than they are under normal circumstances. In the event of an emergency, if no veterinarian is available, you should direct the client to the nearest veterinary practice or animal hospital where emergency treatment can be provided.

Health Decisions

Only the veterinarian can make decisions affecting the health of a patient. No matter who owns the practice or hospital, only a veterinarian may make medical decisions. Nonveterinary staff members should never diagnose any condition, no matter how sure they are that they know the problem. Treatment should only be provided under the supervision of the veterinarian.

OBEYING THE LAWS

It is illegal and unethical for a veterinarian or his or her staff to break the laws of the **jurisdiction** in which they practice veterinary medicine, or to defraud clients or be deceitful in any other way. No matter what your role is as a member of the veterinary staff, you should always be honest and fair in your professional relations.

Misrepresentation

Veterinarians should never identify themselves as members of any organization by which they have not been certified, including the AVMA. It is also unethical for a veterinary technician, veterinary assistant, practice manager, or other members of the veterinary staff to misrepresent themselves as veterinarians.

Lending Credibility to the Illegal Practice of Veterinary Medicine

It is unethical for a veterinarian or any member of his or her staff to promote an illegal or unethical product or service offered by a nonprofessional organization, group or individual. If you are aware of any illegal activities, including the practice of veterinary medicine by someone other than a licensed veterinarian, you should contact the appropriate authorities.

RESPONSIBILITY TO THE VETERINARY PROFESSION

Duty to the Veterinary Image

Veterinarians and their staff members have a responsibility to enhance and uphold the positive image of practitioners of veterinary medicine in the eyes of colleagues, clients, and the general public. You should always act in an honest, courteous, compassionate, and professional manner, and remember that your appearance and behavior reflect directly upon the veterinarians you work for.

Duty to Professional Colleagues

It is both unethical and unprofessional to **slander** fellow veterinarians, regardless of your personal opinions. All members of the veterinary staff should refrain from making any negative remarks or implications about other veterinarians, veterinary practices, or veterinary hospitals. This also includes making any defamatory comments about any of the veterinarians for whom you work. Conversations regarding any personal disputes with a veterinarian should be held in private, out of earshot of clients, co-workers, and the public.

Enhancing and Improving Veterinary Knowledge and Skills

Veterinarians, just as any other health professionals, should engage in ongoing education in their field and collaborate with their colleagues to share and expand their knowledge. Continuing education is required for veterinary technicians, and is recommended for veterinary assistants. The more you know about animal care and your profession, the better able you will be to serve your patients and clients.

DUTY TO SOCIETY

It is highly recommended that veterinarians and staff members be involved in the community and contribute to public health efforts that relate to veterinary medicine. The members of the veterinary staff should be available and willing to provide assistance as needed when public health is threatened.

CONFIDENTIALITY

All members of the veterinary staff, from the veterinarian to the administrative staff to the cleaning people have the responsibility of protecting client and patient information. Releasing any client information, even something as seemingly harmless as the name of client or patient that comes to your office, can be a violation of this **confidentiality**. Such a breach of confidentiality could also be in violation of the law and result in a lawsuit and the loss of your job.

MAINTAINING A CLEAN MIND, BODY, AND SPIRIT

If a veterinarian is under the influence of alcohol, drugs, or other substances that affect judgment, he should not practice medicine. If you feel a veterinarian has a **substance abuse** problem, you should encourage treatment. If you feel any member of the veterinary staff, including yourself, has a substance abuse problem, encourage or seek out treatment. Attempting to function with impaired judgment is particularly dangerous because the behavior and decisions of every member of the veterinary care team affect the health of patients, and can potentially affect the health of humans.

THE VETERINARIAN-CLIENT-PATIENT RELATIONSHIP

The **veterinarian-client-patient relationship** (VCPR) is the professional relationship between the veterinarian, the client, and the patient. When this relationship is established, it should be understood that the veterinary practice or animal hospital is there to provide appropriate, adequate veterinary health care, and the client should follow the veterinarian's instructions for the patient. It should also be understood that the veterinary staff exists to aid the veterinarian in providing care and to serve the client as directed by the veterinarian. As part of this agreement, the practice must keep and maintain medical records for each patient. The veterinary staff must properly update, file, and protect these records to facilitate optimal care and maintain confidentiality. The information must be kept confidential unless the owner of the patient consents to its release or a court orders the release of the information. While these medical records concern the patient, they are the property of the practice. However, said records should never be copied, distributed, or removed from the practice without the knowledge and consent of the patient's owner.

Prescriptions

A VCPR is required for a veterinarian to prescribe medication. Veterinarians should never make out a prescription for a nonpatient, nor should any member of the veterinary staff dispense prescription medication, especially controlled substances, without a prescription or to a nonpatient. Only a veterinarian may write out a prescription.

Termination of the VCPR

If there is no ongoing medical condition, the veterinary practice or animal hospital must notify the client that it no longer wishes to serve the patient and client, if it wants to terminate the relationship. If there is an ongoing medical or surgical condition, the VCPR can only be terminated if the client is directed to another veterinarian for care. The veterinary practice or animal hospital must provide care until the transition is complete. Clients may terminate the relationship at any time for any reason.

SUMMARY

As medical practitioners, veterinarians must adhere to a strict code of ethics. So too must the members of the veterinary staff, including everyone from the receptionist to the veterinary assistant to the veterinary technician. Each person should behave in a professional manner, and not allow any outside influences to affect the way the practice or hospital is run. The clients and patients of a veterinary hospital or practice are due the same respect and confidentiality that people are due in human health care.

CASE STUDY

Jim, a veterinary assistant, was sorting the mail at the front desk when a woman burst through the front door, carrying a small dog that was bleeding badly.

"You have to help me, my dog is hurt!"

Rushing to the front door, Jim said, "Ma'am, please try to calm down and tell me what happened."

"I was walking my dog at the park, when this other big dog came along and they started fighting!" she replied. "We were able to get away, but he's hurt . . . we need to see a vet NOW!"

"She's bleeding pretty badly," Jim said, handing the woman a towel. "Use this to apply pressure on the wound. She's going to need stitches, and probably a tetanus shot."

"Okay, but I need to see a vet!"

"I'm sorry, but Dr. Howard is out sick today, Dr. Fein is at lunch, and Dr. DeRita is out on a farm call. I'm writing down directions to the emergency veterinary clinic in the next town over, and I'll call ahead to let them know you're coming. They're not as good as our doctors, but they should be able to help you."

"Can't you do anything here? She might die if I can't get there in time!"

"Ma'am, there's nothing I can . . ."

"*Do something!*"

"Ma'am, I cannot help you. Your dog needs to see a veterinarian, and there isn't one here right now. I strongly recommend you continue applying pressure to that wound and take your dog to the emergency clinic."

"I can't believe you won't help me!" she replied, walking out the door. "Fine, I'll go somewhere where people care!"

- Which aspects of Jim's behavior were ethical?
- What should Jim not have done?
- What could Jim have done to handle the situation in a more ethical manner?

REVIEW

Indicate whether the following statements are true or false.

1. Veterinarians may only provide emergency treatment to their own patients.

2. Veterinary technicians should never identify themselves as veterinarians.

3. Veterinarians are morally required to treat each and every patient that requests medical care.

4. It is acceptable to question the ability of other veterinarians, but not the veterinarians that you work for.

5. Veterinarians, veterinary technicians, and veterinary assistants should all take ongoing education courses, attend appropriate seminars, and avail themselves of other educational opportunities.

6. It is acceptable to share patient information because veterinary patients are animals, not people, and cannot get upset.

7. A court order is always required for the release of patient records.

8. Veterinary team members should hold themselves to the same ethical standards as veterinarians.

9. Clients may end the VCPR at any time under any circumstances.

10. Veterinarians may end the VCPR at any time under any circumstances.

ON-LINE RESOURCES

AVMA Principles of Veterinary Medical Ethics

This page contains the full text of the complete 1999 revision of the "Principles of Veterinary Medical Ethics" statement from the American Veterinary Medical Association.

<http://www.avma.org/noah>
Search Term: Medical Ethics

NAVTA Code of Ethics

Ethical standards for veterinary technicians, as suggested by the National Association of Veterinary Technicians in America.

<http://www.navta.net>
Search Term: Code of Ethics

The British Columbia Veterinary Medical Association Code of Ethics

This page contains the most recently updated version of the Code of Ethics from this Canadian professional veterinary organization.

<http://www.bcvma.org>
Search Term: Code of Ethics

PART III

Financial Matters

CHAPTER 9

Fee Collection Procedures, Billing, and Payroll

OBJECTIVES

When you complete this chapter, you should be able to:

- explain how doctors' fees and fee profiles are derived

- recognize various circumstances when credit should or should not be extended

- describe the rules and guidelines involved with extending credit on accounts receivable

- distinguish between legitimate and unadvisable reasons for adjusting or canceling a fee

- prepare billing statements and appropriately worded collection letters

KEY TERMS

encounter form
fee structure
fee profile
cost estimate
credit bureau
credit application
medically indigent
professional courtesy
post
account history
alpha search
International Classification of
    Diseases diagnostic codes

cycle billing
accounts receivable ratio
aging accounts receivable
telephone reminder
written reminder
statute of limitations
Employment Eligibility
    Verification Form
gross pay
salary
net pay
exemptions

Employee's Withholding
    Allowance Certificate
Federal Wage-Bracket
    Withholding Table
Federal Insurance
    Contribution Act
payroll tax
social security number
Wage and Tax Statement
Employer's Quarterly Federal
    Tax Return

INTRODUCTION

In many veterinary practices, one of the veterinary assistant's responsibilities is requesting and accepting client payments. It is important not only to know how to act while doing so, but also how the billing process works, when fees should be adjusted or cancelled, and what to do when an account is past due. Another financial matter you should be aware of is payroll accounting, including what forms you need to fill out, what forms you should expect to receive, and when. Veterinary finances are a sensitive matter that can have far-reaching consequences for each member of the veterinary team if mishandled.

FEE COLLECTION PROCEDURES

While generally the job of the receptionist, collecting payments from customers may be done by the veterinary assistant in certain settings or situations. When this is the case, you attach a charge form or fee statement to the case record before you send the patient's chart into the examining room. Many offices use an **encounter form**, formerly known as a *superbill*, a single form that serves as a charge slip and statement (Figure 9-1). The veterinarian marks the form to indicate the services rendered, and the customer or veterinarian returns the form to you. If the veterinarian did not state different charges, then you simply fill in the predetermined standard fees for the procedures marked off. Give one copy of the bill to the customer as a receipt, and keep the other copy for accounting purposes.

As in any personal contact with customers, you should be polite, friendly, and discreet when requesting and accepting payments. Usually, you will have no problem. However, some customers have other things on their mind after an appointment and forget to stop and pay the bill. You should not assume they are trying to skip payment. Nor should you let the bill go unpaid if your office expects payment when providing services. Simply call to the customer, "Excuse me, Mr. Pan, could you come back for a moment?" Then, privately, say something like, "That will be $40 for today, please," and present the bill. Never shout across the waiting room, "Hey, you forgot to pay!"

Explaining Veterinarians' Fees

Although you accept the money, the veterinarian sets the amount of payment for services rendered. Some veterinary offices follow a strict payment schedule of posted rates that are available to customers. Others have a range of fees that depend on the amount of time spent with the patient and the complexity of the diagnosis and treatment. The **fee structure** depends in large part on the types of services provided. For example, a surgical procedure might be billed by the hour, a vaccination might have a set cost, and treatment of injuries might include a combination of fee structures. When setting fees, the veterinarian must

CITY ANIMAL HOSPITAL
123 ANY ROAD, ANY CITY, STATE 99999

H.L. Nelson, V.M.D.   Tel. 555-0000

M   *Joseph Smith*
Address _____

Date *4/14/2000*   By _____

Prof. Service *Daisy*		*18.00*
Surgery		
Anesthetic		
Laboratory	*HW*	*10.00*
X Ray	*Fecal (meg)*	*7.00*
Fluids		
Hospitalization		
Per Day ____		
Approximate		
____ Days		
ECG		
Medicine		
Hospital Stay		
Home Dispensing	*heartgard*	*13.25*
Boarding		
Per Day ____		
____ Days		
Dental		
Vaccine	*DA2P/Parvo SC*	*10.00*
Worming		
Bathing		

This anticipated cost to you is based on prior similar cases. Fees are based on the difficulty of the procedure and the length of time necessary to complete it plus the materials used. Unexpected difficulties may increase the time involved by as much as 25–50%.
On Acct.

Total   **$58.25**

Cash ☐
Charge ☐   Signature *Joseph Smith*

Owner *Joseph Smith*   Date *4/14/20–*
Address _____   Phone _____

CERTIFICATE OF VACCINATION

This is to certify that on this date, I have vaccinated the described below against:

Name *Daisy*   Sex *♀*
Color & Markings *White*
Breed *Bichon*   Age *3*   Weight *11 lbs*
Vaccination   Vaccination
Tag No.   Ser. No.
Veterinarian *H.L. Nelson*
Next Appointment *4/20–*

☒ CANINE
  ☒ Distemper
  ☐ Distemper/Measles
  ☐ (CAV-2)
Hepatitis
  ☒ Lepto C & I
  ☐ Parainfluenza
  ☐ ParvoVirus
  ☐ Bordetella
  ☐ Coronavirus

☐ Rabies
☐ Other ____

☐ FELINE
  ☐ Panleukopenia
  ☐ Rhinotrachetitis
  ☐ Calici Virus
  ☐ Rabies
  ☐ Leukemia
  ☐ Chlamydia
  ☐ Other ____

FIGURE 9-1   Encounter form.

also consider the costs of maintaining the office and staff. Local veterinary societies and veterinary practice management firms can provide veterinarians with information about the fees similar practices charge in the same geographical area and across the country. This information is known as a **fee profile**.

In any case, you will have a fee schedule to go by, not only in filling out the customer's bill, but also in answering the questions that customers will ask you about the costs. Customers, understandably, wish to know what their pets' medical treatment will cost them and how the fees are determined. It can be very difficult to talk about money, but someone in the office must be prepared to do so. Many veterinarians habitually avoid discussing fees with customers. Others may quote a fee but send the customer to the veterinary assistant for an explanation. You must have a good sense of how the fees are determined in the office—and a good sense of interpersonal relations—to discuss what can be an awkward subject. When complicated procedures are involved, fee estimates should be provided.

Customers who are wondering about the cost of treatment are not always assertive about asking questions. They are as nervous as anyone else when talking about money. When you are speaking with customers, particularly those new to the office, you can ask a simple, open-ended question such as, "Do you have any questions about any of our office policies?" This presents an opportunity for them to ask about fees if they wish to do so.

On occasion, customers will be unhappy with fees and will not want to talk to you about the issue. Or they might talk your ear off with complaints. You should be patient with these customers and explain the situation to the veterinarian, who will probably want to talk to them in person.

The Costs Behind the Fees

One of the main reasons veterinary staff members sometimes hesitate to ask clients to pay a bill is because they, like the clients, feel the amount is too high. It is important, though, that you understand the costs related to operating a veterinary practice, which directly affect the amount of the bill. Operational costs often include:

- *Equipment and supplies*. Veterinary equipment, from laboratory equipment to animal cages to diagnostic equipment to furniture and supplies, is *expensive*. Radiograph equipment alone costs tens of thousands of dollars, and cages can range from hundreds to thousands of dollars each. Computers, software, and Internet can easily cost a typical practice more than $10,000, and each piece of furniture in the office can run from hundreds to thousands of dollars. Your practice probably even spends hundreds of dollars each month just on cleaning and disinfecting supplies!

- *Mortgage or lease*. The space in which the practice or hospital is located must be paid for, in the form of lease or mortgage payments. If the property is leased, then the rent probably increases on an annual basis, and if the practice owns the building, property and water taxes must be paid. These costs, too, can easily run into tens of thousands of dollars.

- *Staff salaries.* The salaries of the individuals working at a veterinary practice can easily range from $20,000 to $80,000 (or more!), not to mention the cost of benefits such as health insurance, vacation pay, life insurance, and others. Veterinary fees cover the services provided not only by the veterinarian, but also those of technicians, assistants, receptionists, bookkeepers, and the cleaning service. Although staff members certainly enjoy their work, they also like to be paid, and veterinary fees are the primary source of the practice's income.

- *Training and education.* The first thought that probably comes to mind when considering education costs is the veterinarian's cost of paying off his or her college education, but you must remember that continuing education is required for veterinarians and veterinary technicians. There are also additional costs related to training employees on equipment, computer training, seminars, and other educational expenses. Clients come to a veterinary practice for medical expertise, but the knowledge and skills they expect take time and money to develop.

- *Miscellaneous other costs.* There are also numerous other costs that most people probably would not think of, such as magazine subscriptions for the waiting room, advertising costs, telephone bills, malpractice insurance, insurance on the property, and on and on. Someone even has to pay for the box of tissues on the receptionist's desk!

When you consider the many costs incurred by a veterinary practice, the price of a single procedure generally pales in comparison, and it is easier to realize that the practice is not out to overcharge clients, but rather to cover its own costs in order to operate and continue providing veterinary care. Although you will not necessarily need to explain this to the client complaining about a bill, understanding the costs related to running a veterinary practice or hospital helps you to understand the cost of procedures, making it easier for you to request payment when necessary.

Payment Planning

Any time services are rendered before payment is received, credit has been extended. The extent is often just for a few minutes, because many customers pay their bills on the way out of the office. But when the patient's treatment is ongoing, complicated, or especially expensive—such as surgery—credit is necessary to a greater extent, and a deposit for prolonged, expensive procedures is often required.

Many customers have to pay all or a significant portion of these costs out of their own pocket. As a veterinary assistant, you can ease the financial worries of such customers by preparing a **cost estimate** (Figure 9-2) and helping them work out a payment plan. A common method is to make a fairly large down payment to be followed by smaller monthly payments.

SURGICAL COST ESTIMATE

Name of Patient: _____ Date: _____

Procedure: _____

Your pet's surgery has been scheduled for _____ on _____. You and
your pet should report to the veterinary office between the hours of _____ (AM) (PM) and
_____ (AM) (PM).

Although veterinary expenses are seldom welcomed, knowing in advance what expenses to
expect and how to plan for them can lessen the burden. This estimate is prepared to assist you in
budgeting your surgical costs. The estimated cost for your pet's surgery including anesthesia,
medication, and follow-up care is _____.

The estimated duration of your pet's hospital stay is _____ days at $_____ per day.

PLEASE KEEP IN MIND THIS IS ONLY AN ESTIMATE

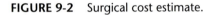

FIGURE 9-2 Surgical cost estimate.

Credit Bureaus. Before extending credit, especially for a significant amount, it might be a good idea to check the customer's credit references. **Credit bureaus** collect information from various sources and provide information for a fee. Veterinarians can check the credit standing of their customers with other creditors. A bureau may contact the office about the status of a customer's account. If a credit bureau is contacting you, that means they have permission from the customer to do so and you may provide certain information from the ledger: the date the account was opened, the current balance, and the highest balance at any time. You should not answer any questions about the customer's character, paying habits, or credit rating. You should not volunteer such information, either.

Credit Applications. Under the Equal Credit Opportunity Act, if you extend credit to one customer, you must extend credit on the same terms to other customers who request it. The only legal reason to deny credit is inability to pay. You can judge ability to pay from the employment and insurance information on the customer registration form, which also serves as a **credit application** in many offices. However, keep in mind that any form used as a credit application may not ask the applicant's sex, race, color, religion, or natural origin. It may *not* ask for the applicant's age (but it may ask for the date of birth). Though it may specify marital status as "married," "unmarried," or "separated," it may not use the terms "divorced," "single," or "widowed."

Truth in Lending. Under the Truth in Lending Act, enforced by the Federal Trade Commission, you must disclose your finance charges to customers

---

WILL YOU TAKE A CHECK?

- A veterinary practice is a business. When accepting checks from customers, use the same precautions as any other business does.

- Make sure the check is signed and completed properly, with the correct date and amount.

- Use discretion in accepting checks from people you do not know. You might want to ask for identification and compare signatures, especially for out-of-town checks.

- With the exception of checks from a customer's insurance carrier, do not accept a third-party check—one that is made out to the customer by someone unknown to you.

- Do not accept a check with corrections on it.

- Do not accept a check with "Payment in Full" written on it, unless of course the entire account really is paid in full. If you accept the check and the amount is less than the total balance of the account at the time, then the customer will not be legally liable to pay the rest of the balance. It is *illegal* to cross out the words "paid in full."

- As soon as you receive a check, stamp the back with the office's deposit endorsement so that nobody else can cash the check if it is lost or stolen. Doing so will remind you to look on the back of the check for notations such as "Payment in Full" that need to be checked out before you accept the check.

- Do not accept postal money orders with more than one endorsement because then your endorsement will be the third, and two is the limit honored by the postal service.

- Avoid accepting any checks for more than the amount due. Some customers will ask you to take a check for more than the amount due so you will give them change. Do not. It makes bookkeeping more complicated than necessary. Customers have even been known to claim that they placed the entire amount of the check on their account. Besides, if the check is returned for insufficient funds, the office is left with not only the loss of the billed amount but also the extra amount given out as cash.

---

with whom you have an agreement to accept payment in four or more installments. You must fill out the disclosure statement (Figure 9-3) even if there are no finance charges. The form must be signed by the customer and kept on file in the veterinary office for two years. This law applies only when you bill the customer each month for the amount of the payment due, not for the full amount owed. When you bill the customer each month for the full amount owed and the customer instead sends in partial payments each month, it is not necessary to complete the disclosure statement.

Credit Cards.   Most veterinary offices and hospitals accept credit cards, not only to give customers more payment options, but also to avoid the sticky problem of extending credit and possibly having to collect delinquent accounts. Credit card companies take up to 8 percent of the amount charged, but veterinary offices may not increase fees for customers who use credit cards.

If your office accepts credit card payments, it probably uses a credit card machine. Like any other piece of office equipment, you should speak to your

---

Sandra Smith, V.M.D.
100 Main Avenue
Anytown
Phone 555-5555

FEDERAL TRUTH IN LENDING STATEMENT
For professional services rendered

Customer _____

Address _____

_____

Patient _____

1. Cash Price (fee for service) $ _____

2. Cash Down Payment $ _____

3. Unpaid Balance of Cash Price $ _____

4. Amount Financed $ _____

5. Finance Charge $ _____

6. Finance Charge Expressed as Annual Percentage Rate $ _____

7. Total of Payments (4 plus 5) $ _____

8. Deferred Payment Price (1 plus 5) $ _____

"Total payment due" (7 above) is payable to _____ at above office address in
_____ monthly installments of $ _____. The first installment is payable on
_____ 20 _____ and each subsequent payment is due on the same day of each
consecutive month until paid in full.

_____   _____
          Date   Signature of Customer

FIGURE 9-3   Federal Truth in Lending statement.

supervisor about being trained in its proper use. Generally speaking, these machines are very easy to use. The typical procedure involves the following steps:

1. *Clear the machine.* You should first make sure the machine is prepared to take a payment. In most cases, this merely requires you to press the "clear" button.
2. *Enter the transaction amount.* When you are sure the machine is ready, you punch in the amount of the charge (being sure that the decimal point is in the correct place – you may need to add two zeros to the end if it is an even dollar amount) and then press "enter." If you make a mistake, you can press the "clear" button and start again.
3. *Enter the credit card number.* Most machines have a "swiper" like those that you see at the register in many grocery stores and depart-

ment stores. These allow you to physically swipe the card through the slot, letting the machine run the card automatically. If the machine will not read the card, you need to enter the information manually. You then type in the credit card number (be sure to verify it again before processing it) and press "enter." You will then be asked for the card's expiration date, which you should key in. Press "enter" again and the machine will process the transaction.

4. *Approval.* If the card is approved, you will get an approval number. If your machine has a built-in printer, it may print it out automatically. If not, you will likely need to use an imprint machine, which allows you to make a carbon imprint of the card on a transaction slip. In this situation, you would take down the approval number on the credit card imprint slip. If the card is not approved, you should follow standard office procedures. Your office may place a call to the credit card company or request a different form of payment. If you are not sure what to do, speak to your supervisor. Regardless of your practice's policy, you should use tact and discretion when discussing a rejected credit card transaction. It does not benefit anyone to embarrass the client.

5. *Signature.* Whether you use a printout or imprint slip, you must get the client's signature on the appropriate line. After signing, the client gets a copy and you file the other copies as directed by your supervisor.

In addition to the credit card machine, some offices are now using computerized credit card programs that allow you to fill in the entries on the computer and then transmit them using a secure Internet connection. Regardless of the method of transaction processing, you must be trained before accepting a credit card payment. Failure to complete a transaction correctly can result in the office not being paid, the client being overcharged, or worse, the wrong person being billed. These are errors you do not want to make.

Adjusting or Canceling Fees

Fees may be adjusted or waived for a variety of reasons. The reason has to be legitimate—fees should never be adjusted as a way to avoid calling in a collection agency to extract payment from customers who can afford to pay. Generally, fees should not be written off when the treatment has had poor results or the patient has died. This may seem heartless at first, but consider the appearances from a legal angle. To some people, writing off the fee in such cases does not look like thoughtfulness and generosity; it looks like an admission of malpractice. However, the veterinarian will occasionally offer a discount to a customer who has disputed the fee (as opposed to disputing the treatment). This offer should be made in writing, using the words "without prejudice" to give the veterinarian the right to reinstate the full charge if the customer does not pay the reduced fee within the specified time limit.

However, if an owner is **medically indigent** (impoverished) or undergoing severe financial hardship, then a collection agency will be no more effective than you are in collecting the amount due. The veterinarian may decide to cancel or reduce the fee in such cases. The information you collect at each customer's first visit to the office will give you a good idea of whether it is necessary to have a frank discussion about the person's ability to pay. This conversation should take place *before* the veterinarian agrees to provide services. Finances must be discussed up front because it should be the veterinarian's choice whether to take on a customer who cannot pay. Most doctors willingly provide service to a limited number of medically indigent animal owners. However, a veterinary office is a business, and no business can afford to give services away to any and all who say they cannot afford to pay. You may find yourself in the situation of having to turn away indigent customers. However, you should never turn them away without first helping them locate other care. Keep a list of the animal shelters in your area that provide assistance to the indigent.

Usually, it is best for the customer and veterinarian to agree on the reduced fee in writing before services are rendered. However, when the customer undergoes hardship in the middle of an animal's treatment, the veterinarian may still choose to reduce the fee. In this case, it is best to enter the full charge on the journal and ledger. Once the customer has paid the agreed-on amount, the rest can be written off as an adjustment. This sort of agreement should also be in writing.

The other common reason for adjusting fees is **professional courtesy**. It has long been a tradition for veterinarians not to charge each other for care. Many veterinarians treat their own employees' animals free of charge, and many offer discounts to veterinary office assistants who work for other doctors. However, there are veterinarians and other medical professionals who feel that professional courtesy is not necessary. It is perfectly ethical to accept payment from those who would rather not accept the professional courtesy.

BILLING CUSTOMERS

Most offices request payment the day services are rendered, and you should follow that policy whenever possible. For routine office visits and relatively inexpensive procedures, it is reasonable to ask for payment at the time of service. But, as we have discussed, it sometimes becomes necessary to extend credit on larger accounts. In these cases, you may be responsible for billing. For some customers, the charge slip is reminder enough. But when payment does not arrive within 30 days, you have to send a statement. You also have to send a statement every 30 days to a customer whose account has a debit or credit balance of one dollar or more.

Every veterinary office has established rules or guidelines for bill collection, including when it is acceptable to extend credit, when payments are due, how the customer is billed, what procedures are followed in collecting delinquent accounts, and how long a delinquent account will be carried before it is turned over to a collection agency.

Computerized Billing

Under traditional accounting methods, the office would keep a daily journal with a page for each day of the year and also pages to summarize monthly and annual totals. Each day's page would have a separate column for the time, the name of the customer and the patient, the service rendered, the charge, and the amount paid (Figure 9-4). The daily journal would be used not just for cus-

May 25, 20—

APPT.	NAME	TYPE OF SERVICE RENDERED		CHARGE		PAID	
9:00	Jones, John/Freckles	90030*		30	00		
9:20	Smith, Beatrice/Pudgy	90050		40	00		
9:40	Hurt, Molly/Chi Chi	Post-Operation Dressing		n/c			
10:00	Scott, Joseph/Ariel	90030		30	00	30	00
10:20	Johnson, Larry/Max	Follow-Up		n/c			
10:40	Taylor, Mary/Whitie	90050		40	00	40	00
11:00	Schultz, Sally/Freeway	90730		30	00		
11:20	Lawrence, Bill/Rex	81000		20	00	20	00
11:40	Levine, Fred/Sandy	90030		30	00		
12:00							
12:20							
12:40							
1:00	Jameson, Sandra/Bitsy	Follow-Up		n/c			
1:20							
1:40	Enright, Kevin/Tike	90015		70	00	70	00
2:00							
2:20	Nuttle, Dan/Fluffy	90015		70	00	70	00
2:40							
3:00	Stevens, Ginny/Leo	90020		120	00		
3:20							
3:40							
4:00	Quilty, Clare/Simba	90000		50	00		
4:20	McManus, Jake/Socks	90020		120	00	120	00
4:40							
5:00							

*The asterisk next to this number denotes a service number; these numbers are essentially codes the veterinary office uses for common procedures.

FIGURE 9-4   A single-entry accounting system uses a daily journal to record financial transactions when services are rendered.

tomers who are paying on their way out of the examination room, but also to record any payments that came in the mail or were dropped off in person to settle an outstanding (unpaid) balance. At the end of the day, the figures would be totaled and entered on the monthly summary (Figure 9-5), and at the end of each month, those figures would be totaled and carried over to the annual summary (Figure 9-6).

At some point in the day, the veterinary assistant would **post** (transfer) each transaction from the daily journal to the accounts receivable ledger, a collection of accounts with a separate card for each customer. This ledger card or file would then act as the client's record of past transactions with the veterinary office and show the current balance of the account.

Thanks to the proliferation of computers and easy-to-use accounting software, it is now much simpler to maintain accounting records and prepare bills and statements. Instead of posting the journal to ledger cards, you type the information into computer files. When filed by computer, the customer ledger is called an **account history**. The account history should have all the same information that you would collect from a new customer: name, address, pet's name, and so on.

The big benefit of computer accounting is its versatility. Software has been designed to perform almost every financial function. Place all charges and payments in the computerized ledger, and you can have a current balance at the touch of a button. Although the computer may contain a customer's complete financial records along with his or her pet's clinical records, the information can be set up so that you can print out only the information you want, not the whole file. Computer software programs can even prepare charge slips (Figure 9-7).

There are several ways in which you can locate a customer's file. Some programs access files by account number, and you just type in the number from an alphabetized cross-referenced listing. If you know whose file you are after, the easiest method is the **alpha search**, or typing in the first few letters of the customer's last name.

On the screen you will see a list of names beginning with those letters, and you simply select the right name. When the file appears on the screen, you can make whatever changes you need to and print out the bill. Computer programs can age accounts and sort them so that statements can be prepared for only those accounts you wish to bill at the time, with a minimum amount of time and effort. These programs also itemize the portions of the account that are current, 30 days past due, over 60 days past due, and so on.

One of the reasons a computer can work so fast is that it is programmed to accept brief codes. For instance, you do not type in the long and sometimes awkward names of diseases; you type in the brief **International Classification of Diseases (ICD) diagnostic codes**. The computer reads the code and automatically writes out the diagnosis. Brief codes can also be programmed to indicate the method of payment (cash, check), any adjustments made to the account, and many other details commonly found in accounting documents.

SUMMARY FOR MONTH _May_ YEAR _20–_

DAY OF MONTH	CHARGES COLUMN (A)		PAYMENTS COLUMN (B)		MISCELLANEOUS SUMMARIES						
1	270	00	100	00							
2	–	–	150	00							
3	–	–	120	00							
4	–	–	–	–							
5	–	–	–	–							
6	301	75	300	00							
7	725	00	150	00							
8	285	50	200	00							
9	–	–	–	–							
10	–	–	–	–							
11	–	–	–	–							
12	65	00	–	–							
13	–	–	–	–							
14	415	50	–	–							
15	230	00	–	–							
16	125	00	–	–							
17	–	–	–	–							
18	–	–	–	–							
19	–	–	–	–							
20	20	00	–	–							
21	377	00	–	–							
22	–	–	–	–							
23	–	–	–	–							
24	–	–	–	–							
25	650	00	350	00							
26											
27											
28											
29											
30											
31											
TOTAL FOR MONTH											
BROUGHT FORWARD	45,540	25	449,21	75							
GRAND TOTAL											

SUMMARY OF EXPENSE (From Reverse Side)		
		AMOUNT
DRUGS AND PROFESSIONAL SUPPLIES		
LAB EXPENSE		
SALARIES		
OFFICE RENT AND MAINTENANCE		
LAUNDRY SERVICE		
ELECTRICITY, GAS, WATER		
TELEPHONE		
DUES AND MEETINGS		
OFFICE EXPENSES (SUPPLIES, ETC.)		
PROFESSIONAL INSURANCE		
BUSINESS TAXES		
INTEREST PAID		
TRAVEL AND ENTERTAINMENT		
TOTAL FOR PRESENT MONTH		
FORWARDED FROM PREVIOUS MONTH	45,182	31
GRAND TOTAL		
MONTHLY BALANCES		
FOR THE PRESENT MONTH		
TOTAL RECEIPTS (COL B)		
TOTAL EXPENSE		
NET EARNINGS		
FOR THE YEAR-TO-DATE		
GRAND TOTAL RECEIPTS		
GRAND TOTAL EXPENSE		
NET EARNINGS		
ACCOUNTS RECEIVABLE (FROM LAST DAY SHEET OF THE MONTH)		
$		CHECKED BY

FIGURE 9-5 Monthly Summary.

ANNUAL SUMMARY

MO.	TOTAL BUSINESS	TOTAL INCOME	TOTAL DISBURSEMENT	DISTRIBUTION OF DEDUCTIBLE DISBURSEMENTS														
				OFFICE EXPENSE	WAGES	AUTO & TRAVEL	MAINTEN-ANCE	MISC. EXPENSE	DRUGS & SUPPLIES	LAB.	RENT	TAXES	INSTRU-MENTS	INTEREST FEES	PROF. FEES	PERSONAL	EQUIP-MENT
JAN	4,375 00	3,750 00	4,111 70	112 75	440 00		72 50	27 50	52 20	110 00	375 00	2,500 00	135 00		45 00	93 50	148 25
FEB.	4,020 00	2,869 00	1,656 25	93 50	440 00		37 75	66 50	48 50	205 00	375 00				60 00	111 20	278 80
MAR.	4,560 00	4,775 00	1,534 00	86 25	550 00		89 00	54 25	30 00	117 00	375 00					87 50	145 00
APR.	3,875 00	4,660 00	5,292 85	81 25	440 00		64 35	178 00	65 00	83 50	375 00	3,517 50	87 25		75 00	228 00	98 00
MAY	3,950 00	3,300 00	1,571 65	75 50	440 00		125 25	28 00	55 50	75 00	375 00					92 40	305 00
JUNE	3,765 00	3,440 00	4,696 50	125 00	550 00		57 50	36 50	37 50	80 00	375 00	3,250 00				185 00	
JULY	1,550 00	2,660 00	1,204 80	94 30	440 00		39 50	45 75	75 25	– 00	375 00				75 00	60 00	
AUG.	2,680 00	1,500 00	1,203 00	97 75	440 00		45 75	68 50	40 00	– 00	375 00					58 00	78 00
SEPT.	4,250 00	3,990 00	4,863 25	76 00	550 00		98 00	77 50	35 50	150 00	375 00	3,250 00	58 00			108 00	85 25
OCT.	4,965 00	4,885 00	1,575 25	87 50	440 00		72 25	47 25	69 00	185 00	375 00		76 50			132 75	90 00
NOV.	4,880 00	4,050 00	1,344 35	89 35	440 00		88 50	42 25	56 75	60 00	375 00					81 00	111 50
DEC.	4,100 00	4,010 00	1,438 25	135 50	550 00		105 25	62 50	50 00	40 00	375 00					120 00	
TOTALS	46,970 00	43,880 00	30,491 85	1154 65	5,720 00		895 60	734 50	615 20	1,105 50	4,500 00	12,517 50	356 75		255 00	1,357 35	1,279 80

Other Income	1,600	00
Total Income	45,480	00
Less nondeductible disbursements	2,637	15
Total deductible disbursements for practice line AA minus line DD	27,854	70
Other than practice deductible disbursements, col 28, line L, last month dist. sheet	–	–
Total deductible disbursements for the year (personal & practice) add lines EE and FF	27,854	70

FIGURE 9-6 Annual summary.

Date	Diagnosis	*	RVS/CPT	Description	Doctor	Amount	Patient Name	
							Account Number	**Statement Date**
3/30/–	4779	11	90060	OFFICE VISIT - EST. PT.	MM	51.00		4/15/–
3/30/–	4779	11	8502424	CBC	MM	23.00	12345	
3/30/–	4779	11	8565124	WESTERGREN SED RATE	MM	14.25		
3/30/–	2724	11	8006224	CORONARY RISK PANEL	MM	51.00	Return this portion with your payment in the envelope provided.	
3/30/–	2724	11	8005824	HEPATIC FUNCTION PANEL	MM	57.00	Make your check payable to:	
3/30/–	7851	11	84444324	TSH	MM	36.75		
3/30/–	7851	11	36415	LABORATORY DRAWING	MM	6.00		
12345	INSURANCE BILLED						IRS # 00-00000000	
	FOR DATE RANGE 301–331 (FOR THE AMOUNT OF $239.00)							
							Check Number	

	Previously Billed Charges Now Past Due				
Current Charges	**Over 30 Days**	**Over 60 Days**	**Over 90 Days**	**Over 120 Days**	**Balance Due**	**Amount Enclosed**
$239.00	$0.00	$0.00	$0.00	$0.00	$239.00	$239.00

Account Number	**Patient Name**	**Billed to:**	**Change of Address: Note Below**

Statement

Patient Name

FIGURE 9-7 Computer-generated charge slip.

With computers, one aspect of human error is eliminated—math mistakes—but you do have to be extra careful not to make any typing errors. You should proofread on screen before saving an entry. Mistakes found later can still be corrected, and the original entry is retained as a safeguard against theft (so nobody can enter a transaction, then delete it and keep the money).

Sending Statements

Customers are more likely to pay promptly if they receive statements at the same time each month. Also, the Fair Credit Billing Act states that the date of mailing statements for credit accounts cannot vary by more than five days unless the debtor has been notified about the change. Since many people pay their bills at the beginning of the month, it is a good idea to mail bills by the 25th of the month, though a lot of offices mail bills on the last day of the month. If you have many statements to send, you might not have time to prepare them all at the same time. You can still maintain the same cycle each month, though. With **cycle billing**, you group accounts alphabetically and divide them into the number of mailings you wish to make. For instance, if you set aside 2 days a month for billing, you could send statements to customers with names beginning with A through M on the 10th, and N through Z on the 25th. Or, if billing 3 days a month, you could mail A through F around the 10th of the month, G through M around the 20th, and N through Z the last day of the month. Some offices even divide accounts into 20 groups and bill a few accounts each day instead of taking a few days to bill large groups of accounts.

Collecting Past Due Accounts

A certain percentage of customers fail to pay their accounts on time. Some are simply negligent; they lose track of time and let their bills go until they get a reminder or two that the account is delinquent. Some are unable to pay because of financial difficulties. And a few—maybe 4 percent—just plain are not willing to pay. A veterinary office is doing pretty well if its **accounts receivable ratio** is less than 2 months. That means that the total of all accounts receivable should equal no more than 2 months' worth of gross charges. (However, in offices that extend credit extensively, a ratio of up to 3 months may be acceptable.) When the ratio gets higher than desirable, it is time to become more aggressive about account collection.

You can compute the accounts receivable ratio using the following formula:

Accounts Receivable Ratio = Current Accounts Receivable
Balance ÷ Average Gross Monthly Charges

Example:

Annual Gross Charges	$144,000
Average Monthly Charges	12,000
($144,000 ÷ 12 months)	

Current Accounts Receivable Balance $18,000

Accounts Receivable Ratio   1.5 months

($18,000 ÷ $12,000)

So, in this example, the accounts receivable ratio is one and a half months. The average monthly charges are $12,000, and the current outstanding balance is $18,000. If the current accounts receivable were $24,000 (the equivalent of two months of monthly charges), the accounts receivable ratio would be two months ($24,000 ÷ $12,000 = 2).

It is important to try to collect on all accounts receivable for several reasons. Obviously, the veterinary office is a business that must bring in income to pay expenses and continue to provide services. In the end, the patients who pay their bills end up absorbing the costs of those who do not, and that is not fair. Delinquent accounts also negatively affect the clients who owe the money, because they may be embarrassed to bring their pets into the office for medical attention when they really need it.

Aging Accounts Receivable. Analyzing accounts that are past due is called **aging accounts receivable** (Figure 9-8). Aging begins on the first day of billing, which depends on the payment plan. It may be the day of service or the day of the first mailed billing. Always keep track of all accounts receivable

ACCOUNTS RECEIVABLE AGE ANALYSIS

Dr.

Address   Date

| Patient's Name | Total Accounts Receivable | Distribution of Accounts Receivable by Age | | | | Comments |
		1-2-3 Months	4-5-6 Months	7-8-9-10-11-12 Months	Over 1 Year	
Adams	450.00	100.00	350.00			
Brown	50.00	50.00				
Condito	100.00		75.00	25.00		
Dawson	200.00		10.00	150.00	40.00	
Emerson	550.00		550.00			
Fine	42.50	42.50				
Greene	65.00	20.00	45.00			
Horan	325.00	325.00				

FIGURE 9-8 A form such as this one enables you to analyze accounts receivable at a glance.

so that you can use appropriate collection techniques. Bills that have been out less than 30 days are considered current and require no additional attempts at collection. But a bill that has been out 60 days is past due and merits a *tactful* and *friendly* reminder.

Telephone Reminders.   A telephone call, if made at the right time and in the right manner, is often the most effective method of debt collection. (See the section on collection calls in Chapter 7.)

Written Reminders.   You can also put your reminder in writing—whether it is a sticker or note attached to the statement, a typed or printed-out form letter, or a preprinted collection card with the blanks filled in (Figure 9-9). Whatever you do, *do not* invade the customer's privacy by using a postcard to communicate your message, and *do not* mark the outside of the envelope with a notation such as "overdue notice." When writing a reminder or collection letter, avoid antagonistic wording such as "You *forgot/neglected/failed* to pay your bill" or condescending language such as "I am sure you just *overlooked* this bill." You will find positive examples of wording in Figure 9-9.

Although some experts believe that the personal letter gets better results, others believe the impersonal form letter or collection card is more effective because it makes the customer feel like she is not the only one being singled out. However, just because you are using a form letter or collection card, that does not mean you should use the same one for every customer. There is a difference between someone who has gotten behind on one payment just this once and someone who habitually allows accounts to become delinquent.

When a bill is 90 days past due, you should give a second reminder. Your computer program can probably be set to alert you when invoices have hit 60, 90, or 90+ days. Most customers, after receiving one or two friendly reminders, will contact you to explain the reason they were not able to pay sooner or to set up a mutually agreeable payment plan. Once an account is past due 90+ days, it has become a collection problem and more persuasive collection techniques must be used.

Forced Collection

It is important to follow up on early friendly reminders with more forceful, though still tactful, reminders. Most offices use a series of three, four, or five letters and/or phone calls that become increasingly firm in their request for payment. After 90 days, it may be office policy to send a "final notice" indicating that unless the account is paid by a specified date, it will be turned over to a collection agency; or, less often, that a lawsuit will be filed in small claims court. Even if you have used the telephone for prior communications, the final notice must be in writing. You should never send out this notice unless the veterinarian has instructed you to do so. And it should not be an idle threat; you must be prepared to follow through on any warning you make. Idle threats may be considered harassment, for which the customer may be able to sue.

Sample 1

M _____ 20 _____

_____

Veterinary bills are payable at the time of service unless special credit arrangements are made.

Please send your check in full or call this office before _____.

Sample 2

M _____ 20 _____

_____

Your account has always been paid promptly in the past, so this overdue account is most unusual. Please accept this note as a friendly reminder of your account due for $ _____.

Sample 3

M _____ 20 _____

_____

Since your pet's care in _____, we have had no word from you about how your pet is feeling or your account due.

If it is impossible for you to pay the full amount of $ _____ at this time, please call this office before _____ so that satisfactory arrangements can be worked out.

Sample 4

M _____ 20 _____

_____

Unless some definite arrangement is made to reduce your balance of $ _____, we can no longer carry your account on our books.

Delinquent accounts are turned over to our collection agency on the 25th of each month.

FIGURE 9-9   A series of four collection letters used by one medical office.

Turning an account over to a collection agency is not a pleasant task, but it is usually the wisest solution—provided that you have given the customer a fair chance to pay the bill or arrange for payment. (As discussed earlier, some bills may be written off or reduced.) But do not wait too long to put delinquent accounts in the hands of professional bill collectors. Consider the law of diminishing returns. Beyond a certain point, the odds of recovering the debt go

---

USING A COLLECTION AGENCY

When the veterinarian decides to turn an account over to a collection agency, you must provide the agency with the following information:

- Debtor's full name
- Name of debtor's spouse or other person responsible for the bill
- Last known address
- Full amount of debt
- Date of last credit or debit on ledger card

Mark the customer's ledger card to indicate that the account has been turned over to the collection agency, and make no further attempts to collect it yourself. If the customer contacts you about the account, refer her to the collection agency.

---

lower and lower, and the time you put into debt collection will not be worth what you get out of it. A reputable collection agency may be able to collect on a higher percentage of these accounts than you can. The agency's 20 percent to 50 percent commission may well pay for itself in the time it saves you.

When an account is considered a 100 percent loss and all other methods have failed, yet the customer can clearly afford to pay, the veterinarian may decide to take legal action. However, litigation is so costly and time consuming that most veterinarians are more likely to write off the account as a bad debt than to take the customer to court. Any legal recourse must be taken within the **statute of limitations**, the time limitation dictated by state law.

Finally, this point is so important that it bears repeating: Your employer has specific policies regarding bill collection. *Always follow office policy and protocol for billing and collections.*

PAYROLL ACCOUNTING

Wage and salary records provide employers with information pertaining to a significant part of operating expenses. Accurate payroll records are necessary for them to calculate the federal, state, and local tax obligations of the business. Employers must also withhold a certain amount of taxes from employees' gross earnings to comply with tax laws, and this process requires quite a bit of paperwork.

Form I-9

Before they can be hired officially, all employees must complete the **Employment Eligibility Verification Form**, or Form I-9, to ensure that they are

U.S. Department of Justice	OMB No. 1115-0136
Immigration and Naturalization Service	**Employment Eligibility Verification**

Please read instructions carefully before completing this form. The instructions must be available during completion of this form. ANTI-DISCRIMINATION NOTICE: It is illegal to discriminate against work eligible individuals. Employers CANNOT specify which document(s) they will accept from an employee. The refusal to hire an individual because of a future expiration date may also constitute illegal discrimination.

Section 1. Employee Information and Verification. To be completed and signed by employee at the time employment begins.

Print Name: Last	First	Middle Initial	Maiden Name

Address (Street Name and Number)	Apt. #	Date of Birth (month/day/year)

City	State	Zip Code	Social Security #

I am aware that federal law provides for imprisonment and/or fines for false statements or use of false documents in connection with the completion of this form.

I attest, under penalty of perjury, that I am (check one of the following):
☐ A citizen or national of the United States
☐ A Lawful Permanent Resident (Alien # A_____)
☐ An alien authorized to work until ___/___/___
(Alien # or Admission #) _____

Employee's Signature	Date (month/day/year)

FIGURE 9-10   Employment Eligibility Verification, Form I-9.

authorized to work in the United States (Figure 9-10). You must submit two proofs of eligibility (such as a valid driver's license and a social security card) to your employer, who will then fill out the remainder of the form and file it accordingly.

Employee Compensation

The Fair Labor Standards Act, or Wages and Hours law, requires employers to maintain comprehensive records for each employee, including the employee's name, social security number, address, class or occupation, rate of pay, basis for wage payment, hours worked, wages at regular rate, overtime premium, additions to or deductions from wages, net payment, and date paid. Many veterinary offices use readily available standard forms for individual employee payroll records (Figure 9-11).

Gross Pay.   **Gross pay** is the total amount earned before payroll deductions. The simplest method of determining an employee's gross earnings is to predetermine and establish a certain rate of **salary** earned per month or per year. But the most frequently used method of determining gross earnings is by the *hourly rate*. Gross pay equals the amount earned per hour multiplied by the number of hours worked. The Fair Labor Standards Act sets minimum hourly wages. By law, a minimum rate of one and one-half times the regular hourly rate must be paid for all hours worked in excess of 40 hours per week. Higher rates, often twice the base rate, can also be paid as a premium for working at night, on Sundays or holidays, or at other less desirable times. (Overtime or premium pay requirements do not apply to the employees on a straight salary, however.)

EMPLOYEE'S EARNINGS RECORD

NAME Mary Crane
ADDRESS 810 Columbia Street
Newport Beach, CA 92663

SOC. SEC. No. 000-00-0000
EMPLOYEE NO. 3

DATE EMPLOYED 8-16-2000 DATE OF TERMINATION
DATE OF BIRTH 7-2-72
MARRIED X SINGLE NUMBER PAY
MALE FEMALE X ALLOWANCES 1 RATE 6.00/hr.
OCCUPATION Clerk
CLOCK NO. 6123

Pay Period Ended	Total Hours	Regular Pay	Overtime	Gross Pay	Cumulative Pay	FICA Taxes	Fed. Income Taxes	Hosp. Ins.	Other Deductions	Total Deductions	Net Pay	Check. No.
1/6	40	240.00		240.00	240.00	18.36	26.40	6.25		51.01	188.99	46
1/13	40	240.00		240.00	480.00	18.36	26.40	6.25		51.01	188.99	59
1/20	40	240.00		240.00	720.00	18.36	26.40	6.25		51.01	188.99	72
1/27	48	240.00	54.00	294.00	1,014.00	22.49	36.90	6.25		65.64	228.36	85
2/4	40	240.00		240.00	1,254.00	18.36	26.40	6.25		51.01	188.99	98
2/11	40	240.00		240.00	1,494.00	18.36	26.40	6.25		51.01	188.99	111
2/18	40	240.00		240.00	1,734.00	18.36	26.40	6.25		51.01	188.99	124
2/25	40	240.00		240.00	1,974.00	18.36	26.40	6.25		51.01	188.99	137
4/25	40	248.00*		248.00*	3,902.00	18.97	26.40	6.25		51.01	196.38	372

*Rate Increased to $6.20/hr.

FIGURE 9-11 Employment payroll record.

Net Pay. **Net pay** is the amount received after payroll deductions, which may include agreed-on deductions such as union dues, insurance premiums, and voluntary savings or investment plans as well as income taxes.

Taxes Withheld

Each week, a certain portion of your wages will be withheld from your check to go toward your income taxes. At the end of the year, this total amount withheld will be reported on a W-2 form, which you will submit with your income tax return. If too much money has been withheld throughout the year, you will receive a tax refund. If not enough has been withheld, you will need to send payment to make up the difference.

Federal Income Tax Withholding. The amount of federal tax income withheld from gross wages depends on the number of personal withholding allowances the employee has claimed. The most common of these withholding allowances are **exemptions** claimed for an employee and an employee's spouse and dependents. For each exemption claimed, a certain amount of an employee's yearly gross wages is exempt from income tax. An additional exemption can also be claimed for a dependent spouse who is 65 or older, or if the employee or dependent spouse is blind. To document the number of exemptions claimed, the employee must fill out an **Employee's Withholding Allowance Certificate**, better known as the Form W-4 (Figure 9-12).

The amount of federal income tax withheld from an employee's gross wages is determined by the **Federal Wage-Bracket Withholding Table** (Figure 9-13). Separate tables are available for weekly, biweekly, semimonthly,

Form **W-4**
Department of the Treasury
Internal Revenue Service

Employee's Withholding Allowance Certificate

► **For Privacy Act and Paperwork Reduction Act Notice, see page 2.**

OMB No. 1545-0010

1 Type or print your first name and middle initial | Last name | 2 Your social security number

Home address (number and street or rural route)

3 ☐ Single ☐ Married ☐ Married, but withhold at higher Single rate.
Note: *If married, but legally separated, or spouse is a nonresident alien, check the "Single" box.*

City or town, state, and ZIP code

4 **If your last name differs from that shown on your social security card, check here. You must call 1-800-772-1213 for a new card.** ► ☐

5 Total number of allowances you are claiming (from line **H** above **or** from the applicable worksheet on page 2) 5

6 Additional amount, if any, you want withheld from each paycheck . . . . 6 $

7 I claim exemption from withholding for 2003, and I certify that I meet **both** of the following conditions for exemption:
- Last year I had a right to a refund of **all** Federal income tax withheld because I had **no** tax liability **and**
- This year I expect a refund of **all** Federal income tax withheld because I expect to have **no** tax liability.

If you meet both conditions, write "Exempt" here . . . . ► 7

Under penalties of perjury, I certify that I am entitled to the number of withholding allowances claimed on this certificate, or I am entitled to claim exempt status.
Employee's signature
(Form is not valid
unless you sign it.) ► **Date** ►

8 Employer's name and address (Employer: Complete lines 8 and 10 only if sending to the IRS.) | 9 Office code (optional) | 10 Employer identification number

Cat. No. 10220Q

FIGURE 9-12 Employee's Withholding Allowance Certificate, Form W-4.

MARRIED Persons—**WEEKLY** Payroll Period

If the wages are—		And the number of withholding allowances claimed is—										
At least	But less than	0	1	2	3	4	5	6	7	8	9	10
		The amount of income tax to be withheld is—										
$740	$750	$93	$86	$79	$72	$65	$58	$51	$44	$37	$30	$23
750	760	95	88	81	74	67	60	53	45	38	31	24
760	770	96	89	82	75	68	61	54	47	40	33	26
770	780	98	91	84	77	70	63	56	48	41	34	27
780	790	99	92	85	78	71	64	57	50	43	36	29
790	800	101	94	87	80	73	66	59	51	44	37	30
800	810	102	95	88	81	74	67	60	53	46	39	32
810	820	105	97	90	83	76	69	62	54	47	40	33
820	830	108	98	91	84	77	70	63	56	49	42	35
830	840	111	100	93	86	79	72	65	57	50	43	36
840	850	114	101	94	87	80	73	66	59	52	45	38
850	860	116	103	96	89	82	75	68	60	53	46	39
860	870	119	106	97	90	83	76	69	62	55	48	41
870	880	122	109	99	92	85	78	71	63	56	49	42
880	890	125	112	100	93	86	79	72	65	58	51	44
890	900	128	114	102	95	88	81	74	66	59	52	45
900	910	130	117	104	96	89	82	75	68	61	54	47
910	920	133	120	107	98	91	84	77	69	62	55	48
920	930	136	123	110	99	92	85	78	71	64	57	50
930	940	139	126	112	101	94	87	80	72	65	58	51
940	950	142	128	115	102	95	88	81	74	67	60	53
950	960	144	131	118	105	97	90	83	75	68	61	54
960	970	147	134	121	108	98	91	84	77	70	63	56
970	980	150	137	124	110	100	93	86	78	71	64	57
980	990	153	140	126	113	101	94	87	80	73	66	59
990	1,000	156	142	129	116	103	96	89	81	74	67	60
1,000	1,010	158	145	132	119	106	97	90	83	76	69	62
1,010	1,020	161	148	135	122	108	99	92	84	77	70	63
1,020	1,030	164	151	138	124	111	100	93	86	79	72	65
1,030	1,040	167	154	140	127	114	102	95	87	80	73	66
1,040	1,050	170	156	143	130	117	104	96	89	82	75	68
1,050	1,060	172	159	146	133	120	106	98	90	83	76	69
1,060	1,070	175	162	149	136	122	109	99	92	85	78	71
1,070	1,080	178	165	152	138	125	112	101	93	86	79	72
1,080	1,090	181	168	154	141	128	115	102	95	88	81	74
1,090	1,100	184	170	157	144	131	118	104	96	89	82	75
1,100	1,110	186	173	160	147	134	120	107	98	91	84	77
1,110	1,120	189	176	163	130	136	123	110	99	92	85	78
1,120	1,130	192	179	166	152	139	126	113	101	94	87	80
1,130	1,140	195	182	168	155	142	129	116	102	95	88	81
1,140	1,150	198	184	171	158	145	132	118	105	97	90	83
1,150	1,160	200	187	174	161	148	134	121	108	98	91	84
1,160	1,170	203	190	177	164	150	137	124	111	100	93	86
1,170	1,180	206	193	180	166	153	140	127	114	101	94	87
1,180	1,190	209	196	182	169	156	143	130	116	103	96	89
1,190	1,200	212	198	185	172	159	146	132	119	106	97	90
1,200	1,210	214	201	188	175	162	148	135	122	109	99	92
1,210	1,220	217	204	191	178	164	151	138	125	112	100	93
1,220	1,230	220	207	194	130	167	154	141	128	114	102	95
1,230	1,240	223	210	196	183	170	157	144	130	117	104	96
1,240	1,250	226	212	199	186	173	160	146	133	120	107	98
1,250	1,260	228	215	202	189	176	162	149	136	123	110	99
1,260	1,270	231	218	205	192	178	165	152	139	126	112	101
1,270	1,280	234	221	208	194	181	168	155	142	128	115	102
1,280	1,290	237	224	210	197	184	171	158	144	131	118	105
1,290	1,300	240	226	213	200	187	174	160	147	134	121	108
1,300	1,310	242	229	216	203	190	176	163	150	137	124	110
1,310	1,320	245	232	219	206	192	179	166	153	140	126	113
1,320	1,330	248	235	222	208	195	182	169	156	142	129	116
1,330	1,340	251	238	224	211	198	185	172	158	145	132	119
1,340	1,350	254	240	227	214	201	188	174	161	148	135	122
1,350	1,360	259	243	203	217	204	190	177	164	151	138	124
1,360	1,370	259	246	233	220	206	193	180	167	154	140	127
1,370	1,380	262	249	236	222	209	196	183	170	156	143	130
1,380	1,390	265	252	238	225	212	199	186	172	159	146	133

$1,390 and over

FIGURE 9-13 Federal Weekly Wage-Bracket Withholding Table.

monthly, and miscellaneous pay periods for persons who are married and for persons who are unmarried. To use the table, check the employee's W-4 for the number of exemptions claimed. Then locate the employee's wage bracket in the first two columns of the table, and the amount to be withheld will appear on the line of the wage bracket under the column heading that shows the number of exemptions claimed.

FICA Tax Withholding.   Besides income taxes, employers must withhold taxes levied under a law called the **Federal Insurance Contribution Act** (FICA)—social security taxes. These FICA taxes are imposed, in equal amounts, on *both* covered employers and their employees. The employer's share is called a **payroll tax**. Congress establishes the tax rate, and there is a cap on the amount of wages subject to the FICA tax. Social security benefits are based on the average earnings of the worker during the years of employment. To keep track of gross earnings and FICA taxes withheld, each employee must obtain a nine-digit **social security number** (application forms are available from local Social Security offices, Internal Revenue offices, and post offices).

State and Local Tax Withholding.   In addition to federal income taxes and FICA taxes, employers in most states and cities must also deduct state and city income taxes from their employees' earnings. The amount to be withheld is based on an employee's gross wages and, in some states, claimed exemptions.

Form W-2.   At designated times during the year, the employer must pay all payroll taxes and employee withholdings to the proper agencies. Within one month after the end of a calendar year, the employer must furnish each employee with a **Wage and Tax Statement**, or Form W-2, which shows the amount of wages earned and the amounts of FICA, federal, state, and local income taxes that were withheld during the year (Figure 9-14).

Form 941.   The veterinarian, as employer, must file an **Employer's Quarterly Federal Tax Return**, Form 941 (Figure 9-15). On this form, the employer must furnish information pertaining to total wages subject to withholding, the amount of federal income tax withheld from the employees, the total wages subject to FICA taxes, and the combined amount of employees' and employer's FICA tax.

Unemployment Insurance

The Social Security Act of 1935 provided for temporary benefits for those who become unemployed as a result of economic conditions beyond their control. Unemployment insurance benefits are paid and administered by individual state departments of unemployment. States differ in regard to the types of covered employment and the number of workers an employer must have before the tax is levied. Federal funds are raised by a payroll tax on employers

1 Control Number			OMB No._1545-0008
	222		

2 Employer's name, address, and zip code Robert Barton, V.M.D. 4401 Birch Street Newport Beach, CA 92663		6 Statutory Deceased Pension Legal 942 Subtotal Deferred Void Employee Plan rep. Emp. comp. ☐ ☐ ☐ ☐ ☐ ☐ ☐ ☐	
		7 Allocated tips	8 Advance EIC payment
3 Employer's identification number 24-02060310	4 Employer's state I.D. Number	9 Federal income tax withheld 821.60	10 Wages, tips, other compensation 10,400.00
5 Employee social security number 851-22-4057	11 Social security tax withheld 795.60	12 Social security wages 10,400.00	
19 Employee's name, address, and zip code Mary H. Blatt 22 Elm Street Newport Beach, CA 92663		13 Social security tips	14 Medicare wages and tips
		15 Medicare tax withheld	16 Nonqualified plans
20		21 17	18 Other

24 State income tax 228.80	25 State wages, tips, etc. 10,400.00	26 Name of State California	27 Local Income tax 104.00	28 Local, wages, tips, etc. 10,400.00	29 Name of locality Newport Beach

Copy 1 for State, City, or Local Tax Department   Department of the Treasury—Internal Revenue Service

FIGURE 9-14   Wage and Tax Statement, Form W-2.

who qualify, but employers generally make larger payments directly to the state agencies that administer the program.

SUMMARY

Collecting payment from patients is an important part of the veterinary assistant's job, and you will need to patient, respectful, and discreet when requesting payment. Oftentimes, you will also need to handle questions and complaints about fees, although the veterinarian sets the fees, not the veterinary assistant. Again, you should always be patient and respectful when answering questions.

When services are provided prior to payment, credit is being extended to the patient. Many veterinary offices offer credit during the office visit, and require payment the same day. For larger bills, the payment process can take a bit longer.

Most offices use computer accounting programs for billing and statements. These programs make it easier to locate, enter, and retrieve information.

Monthly statements should be sent to patients with a balance of one dollar or more, to remind them that payment is due. When a bill is 60 days

Form **941**
(Rev. January 2003)
Department of the Treasury
Internal Revenue Service (99)

Employer's Quarterly Federal Tax Return

▶ See separate instructions revised January 2003 for information on completing this return.

Please type or print.

OMB No. 1545-0029

Enter state code for state in which deposits were made **only** if different from state in address to the right ▶ (see page 2 of separate instructions).

Name (as distinguished from trade name)

Date quarter ended

Trade name, if any

Employer identification number

Address (number and street)

City, state, and ZIP code

T	
FF	
FD	
FP	
I	
T	

If address is different from prior return, check here ▶

IRS Use

| 1 1 | 1 1 1 | 1 1 1 1 1 | 2 | 3 3 3 | 3 3 3 3 3 | 4 4 4 | 5 5 5 |
| 6 | 7 | 8 8 8 8 8 8 8 | 9 9 9 9 9 | 10 10 10 | 10 10 10 10 10 10 |

A If you **do not have to file** returns in the future, check here ▶ ☐ and enter date final wages paid ▶

B If you are a seasonal employer, see **Seasonal employers** on page 1 of the instructions and check here ▶ ☐

1	Number of employees in the pay period that includes March 12th . ▶	1		
2	Total wages and tips, plus other compensation		2	
3	Total income tax withheld from wages, tips, and sick pay		3	
4	Adjustment of withheld income tax for preceding quarters of **this calendar year**		4	
5	Adjusted total of income tax withheld (line 3 as adjusted by line 4)		5	
6	Taxable social security wages	6a	× 12.4% (.124) =	6b	
	Taxable social security tips	6c	× 12.4% (.124) =	6d	
7	Taxable Medicare wages and tips . . .	7a	× 2.9% (.029) =	7b	
8	Total social security and Medicare taxes (add lines 6b, 6d, and 7b). **Check here if wages are not subject to social security and/or Medicare tax** . . . . ▶ ☐		8	
9	Adjustment of social security and Medicare taxes (see instructions for required explanation) Sick Pay $_____ ± Fractions of Cents $_____ ± Other $_____ =		9	
10	Adjusted total of social security and Medicare taxes (line 8 as adjusted by line 9)		10	
11	**Total taxes** (add lines 5 and 10)		11	
12	Advance earned income credit (EIC) payments made to employees (see instructions) . . .		12	
13	Net taxes (subtract line 12 from line 11). **If $2,500 or more, this must equal line 17, column (d) below (or line D of Schedule B (Form 941))**		13	
14	Total deposits for quarter, including overpayment applied from a prior quarter		14	
15	**Balance due** (subtract line 14 from line 13). See instructions		15	
16	**Overpayment.** If line 14 is more than line 13, enter excess here ▶ $_____ and check if to be: ☐ Applied to next return **or** ☐ Refunded.			

- **All filers:** If line 13 is less than $2,500, **do not** complete line 17 or Schedule B (Form 941).
- **Semiweekly schedule depositors:** Complete Schedule B (Form 941) and check here . . . . ▶ ☐
- **Monthly schedule depositors:** Complete line 17, columns (a) through (d), and check here. . . . . . . ▶ ☐

| 17 | **Monthly Summary of Federal Tax Liability.** (Complete **Schedule B (Form 941)** instead, if you were a semiweekly schedule depositor.) | | | |
| **(a)** First month liability | **(b)** Second month liability | **(c)** Third month liability | **(d)** Total liability for quarter |

Third Party Designee

Do you want to allow another person to discuss this return with the IRS (see separate instructions)? ☐ **Yes.** Complete the following. ☐ **No**

Designee's name ▶

Phone no. ▶ ( )

Personal identification number (PIN) ▶

Sign Here

Under penalties of perjury, I declare that I have examined this return, including accompanying schedules and statements, and to the best of my knowledge and belief, it is true, correct, and complete.

Signature ▶

Print Your Name and Title ▶

Date ▶

For Privacy Act and Paperwork Reduction Act Notice, see back of Payment Voucher.

Cat. No. 17001Z

Form **941** (Rev. 1-2003)

FIGURE 9-15 Employer's Quarterly Federal Tax Return, Form 941.

overdue, it is considered past due, so you will need to send some reminders, either over the phone or through the mail, remembering to be respectful, but firm. Bills that are over 90 days past due are considered problem collections and may be turned over to a collection agency, which will take 20 percent to 50 percent of the amount owed. It is also possible to take the person to court, but that can be timely and expensive. In any case, you should be familiar with your office's own billing and collections policies, as they can differ from practice to practice.

CASE STUDY

It was the 25th of the month again, and Calvin was reviewing client accounts to identify any aging accounts receivable. He noticed that Mrs. Pendergast still had not paid the remaining $100 balance she owed for her pet poodle's surgery 2 months ago, so he mailed her a friendly, but firm, reminder of the amount due.

A week later, payment had not arrived, so he gave her a call. Unfortunately he got her answering machine, so he left a message asking her to call back. She never did. Several follow-up calls garnered the same result.

A second statement marked "final notice" and warning that the invoice would be turned over for collection if not paid, went out on the 25th of the next month, and more calls were made, but to no avail. The bill was finally turned over to the collection agency, and within 2 weeks, payment had been made (less 25% for their fee). A week later, Mrs. Pendergast sent an angry letter to the office, complaining of harassment and indicating that she and her pet would not be back.

- Which fee collection guidelines did Calvin follow?
- Was Mrs. Pendergast's complaint of harassment justified?
- What could have been done to prevent this collection problem?

REVIEW

For items 1–9, choose the correct term in each group.

1. Fees are determined by the (veterinary office assistant/veterinarian).

2. An encounter form is designed for (large fees/several purposes).

3. (Written reminders/telephone calls) least invade a customer's privacy in attempting to collect fees due to a veterinarian.

4. You (should/should not) attempt to provide cost estimates for customers who request them.

5. On credit applications you may not ask for the applicant's (date of birth/sex).

6. You must complete a federal Truth in Lending statement only for those customers who (are paying finance charges/have agreed to pay in a certain number of installments).

7. Adjusting a fee is unwise when the customer (has died/is a veterinarian).

8. A medically indigent customer is (chronically ill/impoverished).

9. An unacceptable third-party check is made out to the (veterinarian/customer) by an (insurance company/unknown person).

10. What is cycle billing?

11. What information should you provide to a collection agency?

ON-LINE RESOURCES

Fee Schedule for Veterinary Services

This web page, part of the University of Florida Animal Care Services site, lists various veterinary services offered and the fees charged for each. Reviewing this site can help students to familiarize themselves with the way veterinarians charge for services.

<http://www.health.ufl.edu>
Search Terms: Veterinary Services, Fees

How to Collect Client Debt

This article explains how to collect on past due invoices while keeping clients, ranging from the gentle reminder to collection options, all the way to litigation. Brought to you by Score.org, part of the How-To on-line network.

<http://www.score.org>
Search Terms: Client Debt: Workshops

How to Collect Payments

More tips and hints on collecting payments, from Entrepreneur.com, the on-line version of the popular small business and finance magazine.

<http://www.entrepreneur.com>
Search Terms: Collecting Payments; Your Business

How to Get Paid On Time

Also from Entrepeneur.com, this article provides suggestions and strategies for getting paid on time.

<http://www.entrepreneur.com>
Search Terms: Collecting; Payments; Your Business

Glossary

A

account history The computerized version of the customer ledger, containing client information and transaction history.

accounts receivable ratio The number of current accounts receivable compared to the average gross monthly charge.

addressee The person to whom a letter is addressed.

admitted Formally entered a patient into the care of a veterinary hospital.

aging accounts receivable Analyzing accounts that have a past due balance.

alpha search Searching a computer accounting database by typing in the first few letters of the customer's last name.

alphabetic filing A filing system based on the client's name; each letter of the alphabet generally has its own color code.

American Animal Hospital Association (AAHA) An organization dedicated to professional veterinary development, quality hospital standards, and excellence in the delivery of veterinary medicine.

American Veterinary Assistants Association (AVAA) A membership association of and for veterinary assistants, working together to bring recognition and respect to the veterinary assistant profession.

American Veterinary Medical Association (AVMA) A professional association dedicated to advancing the field of veterinary medicine and all of its related aspects.

anesthesia/surgery logbook Logbook used to record the use of anesthesia during surgery; documents the date, the patient and client, the procedure(s) performed, all drugs administered (including the exact volumes given and the routes of administration), the length of the anesthetic event, the length of the procedure, and the identities of the surgeon(s) and anesthetist.

autoclave A device used to sterilize surgical instruments.

B

body language The gestures and mannerisms that help people communicate.

bookkeeper A person hired by a veterinary practice or hospital to take care of the accounting aspects of the practice, including payroll, paying bills, and ordering supplies.

broadband A high-speed Internet connection that allows another type of signal to travel over the same line simultaneously. Such access is often provided through television cable companies. DSL service provided over the telephone line is also a form of broadband Internet access.

burnout A term used to describe a condition that results from too much stress over an extended period.

C

call board The component of the call director, containing buttons representing the different stations in the office.

call director A telephone unit with stations that allow several calls to come in at once.

capital equipment Equipment that has a fairly long life expectancy and contributes to the generation of practice revenues.

CD-R A CD-ROM that allows a user to write data to the disk if he or she has a CD-write drive. A CD-R can only be written to once. Another variety of CD-ROM, the CD-RW, allows you to overwrite the information over and over again.

CD-ROM Compact Disc-read only memory. An optical storage device that uses a laser to read data. A CD-ROM generally holds 650mb (megabytes) of data. A CD-ROM drive is required to access the information on a CD-ROM.

central processing unit The "brains" of the computer; the box containing the hard drive, motherboard, built-in drives, and such. This is the piece of the computer that everything else connects to.

centrifuge A type of laboratory equipment used to separate the components of blood.

clean as you go   The practice of putting things back where you found them when you are finished using them, tidying up an area after a procedure, clearing counters that have become unnecessarily cluttered, and dusting.

communicated   Exchanged or received information.

communication process   Using the four elements of communication: message, the information to be sent; sender, the person sending the message; channel, the method through which the message is sent; and receiver, the person receiving the message.

confidentiality   Maintaining secrecy regarding a client or patient's health, financial status, and other vital information.

consent form   A written form signed by the client stating that he or she gives his or her informed consent for the procedures to be performed and acknowledges the potential risks and benefits.

controlled substance logbook   Logbook used to record the use of controlled substances in order to help prevent drug abuse and to require accountability of drug use; contains the date, client's name, patient's name, the amount dispensed, the amount remaining, the names of veterinarians authorizing the use of the medication, and the individual actually handling the medication.

conventional format   Another term for the source-oriented medical record; ordering medical records as acquired, in chronological order, and for multiple patients.

corporate veterinary medicine   Veterinary services provided in specialized settings rather than the traditional veterinary practice or animal hospital.

cost estimate   A statement, in writing, explaining the fees that will be charged for procedures, tests, medication, and such, related to medical treatment. The cost estimate is provided before treatment is given.

courtesy   Putting the needs of other people above your own.

credit application   A form filled out by a client seeking services with the understanding that the client will pay for them later. Such applications generally inquire about employment history, current and past creditors, other forms of income, and other information.

credit bureau   An organization that collects credit information about individuals and then provides reports to potential creditors, with the approval of the person involved.

cross-training   Learning other people's duties to the level that you can perform their jobs if needed.

cycle billing   A billing method in which accounts are grouped medically and then divided into the number of mailings you wish to make. Each group of bills is sent out on a predetermined date in such a manner that all bills go out each month.

D

database   A collection of information organized into easily searchable fields.

defense mechanisms   Adjustments we make in our behavior, usually unconsciously, to help us deal with the experiences or feelings that cause us psychological stress.

denial   Refusing to admit or acknowledge that a traumatic, stressful situation exists in order to avoid dealing with it.

digital camera station   The docking port for a digital camera; the device that a digital camera is placed in to transmit data from the camera to the computer.

discharge   To formally release a patient from the care of a veterinary hospital.

displacement   Transferring emotions about one person, idea, or situation to another, more acceptable or easier target.

E

empathy   Being able to feel and understand what the other person is feeling.

Employee's Withholding Allowance Certificate   Form W-4; the form documenting the number of exemptions claimed by an individual.

Employer's Quarterly Federal Tax Return   Form 941; a tax return that must be filed by the employer four times a year, stating total wages paid and the amount of taxes withheld.

Employment Eligibility Verification Form   Form I-9; the form that must be filled out and filed with the government to ensure that the person is authorized to work in the United States.

encounter form   Formerly known as a *superbill*, an encounter form contains basic client information and spaces to indicate procedures performed on the day of the visit. The encounter form is attached to the patient's chart before seeing the doctor, then is filled out by the veterinarian, and finally is returned to the receptionist or veterinary assistant before leaving.

enunciation   The way a person forms or articulates words.

establishing the matrix   The advanced preparation of the appointment book.

ethics   Rules of moral conduct.

euthanasia   The practice of ending an animal's life humanely and painlessly when it is suffering from a painful, incurable disease or condition.

exemptions   Withholding allowances, generally for the employee's spouse and children.

extranet   A computer network that allows limited access to off-site users. A user name and password is required to gain access, as a firewall is used to block unauthorized users. Off-site users generally access the extranet by using a modem to dial into the server.

F

Federal Insurance Contribution Act   Commonly known as FICA; the law that requires taxes to be paid by both the employer and the employee for social security.

Federal Wage-Bracket Withholding Table   Special chart used to determine the amount of federal tax withholdings to be subtracted from an employee's paycheck.

fee profile   A breakdown of fees generally charged by veterinary practices and hospitals for similar services in a particular geographical area. The fee profile can be used to determine how much to charge clients.

fee structure   The way a practice charges for its services. The fee structure may be a flat fee per procedure, based on the amount of time spent with the client, or a combination of these and possibly other methods.

feedback   The return message in the communication process.

firewall   A device or program used to limit access to a server.

fixed office hours   A scheduling practice in which office hours are set, and rather than making appointments, clients and patients are seen on a first-come, first-served basis.

floppy disk   A small storage device in disk form. Most floppy disks used today are the 3½″ style that hold up to 1.44mb (megabytes) of data.

flow scheduling   Scheduling patients for segments, based on how long a typical appointment lasts.

G

gross pay   The amount earned before payroll deductions.

H

hardware   The physical components of a computer.

health certificate   A special certificate required for interstate and international transport of an animal; generally only filled out by the veterinarian.

I

intellectualization   Using reasoning to avoid the truth, as a way of denying strong feelings that may be socially unacceptable or difficult to accept.

International Classification of Diseases diagnostic codes   The industry standard classification of ailments by numerical code; used for account billing in place of typing in the entire name of the ailment.

Internet   The global network of computer networks that allows millions of computer users to exchange information. There is no central server or single primary Internet source. Each individual server on the Internet can be used to send and receive information. Access to the Internet is generally obtained through an Internet service provider.

Internet service provider   A company that provides Internet access through its servers. Most offer monthly subscriptions and provide a user name and password that allow you to browse the World Wide Web and Usenet, and send and receive e-mail.

interpersonal communication skills   The way you communicate with other people.

intranet   An internal computer network typically used in a single location. A user name and password is required to access the intranet, and a firewall is used to keep any unauthorized users out.

inventory   The products and supplies a veterinary practice has on hand.

J

jurisdiction   The area in which a governing body has the legal power to administer and enforce the law.

K

keyboard   An input device that allows you to enter letters, numbers, and symbols into various computer programs.

kindness   Being helpful, compassionate, and friendly, and treating others how you would want to be treated if the situation were reversed.

L

laboratory logbook   Provides a synopsis of the laboratory tests used in the care of animal patients; contains at the very least, the date, client and patient names, tests performed, and whether the tests are run in-house.

listening   Paying attention to what is being said, how it is being said, and the nonverbal messages (including body language) being sent.

logbook   Detailed records on patients, clients, and procedures relating to x rays, anesthesia/surgery, and controlled substances.

M

malingering   Deliberately pretending to be sick in order to avoid a situation that causes anxiety.

master problem list   A general health record indicating any problems the patient is experiencing, as reported by the client or observed by a member of the veterinary team; contains dates of inoculation, results of heartworm tests, drug allergies, ongoing medical conditions, and so forth.

material safety data sheets (MSDS)   Published by the manufacturers of various products used by the practice, these sheets convey special handling requirements and user restrictions.

medically indigent   An impoverished individual unable to pay for medical care.

medical records   Records that serve as a detailed description of medical issues, their progress, and their resolution; they are used as benchmarks for measuring improvement or deterioration of an animal that has a medical problem.

medical supplies   Items used in the everyday functioning of the practice that help doctors and technicians do their jobs.

message   The information sent or received in communication.

modem   A communications device that allows the computer user to connect to another computer or network over a telephone or cable line, to transfer data.

monitor (noun)   The computer screen.

monitor (verb)   To keeping track of the clients and patients in the waiting area, knowing who is there and why they have come to the practice.

monotone   Not showing a change in feeling or pitch while speaking.

mouse   The handheld pointer device used to manipulate the cursor or pointer on the screen.

N

National Association of Veterinary Technicians in America (NAVTA)   An organization of veterinary technicians dedicated to fostering high standards of veterinary care, promoting the health care team, and advancing the veterinary technology profession.

National Institute for Occupational Safety and Health (NIOSH)   The federal agency responsible for conducting research and making recommendations for the prevention of work-related disease and injury.

net pay   The amount of earnings received after payroll deductions.

network   Two or more computers connected to one another.

neuter/spay certificate   A certificate that allows the client to prove that an animal is no longer sexually intact.

numerical filing   A filing system based on each client's or patient's assigned number, rather than the name.

O

Occupational Safety and Health Administration (OSHA)   Provides guidelines created by the U.S. Department of Labor for safe and healthful working conditions.

objective information   Factual, measurable data, such as an animal's weight.

operating system   A special kind of computer program that runs the computer and all of its other software applications. The most common operating systems are Microsoft Windows and Mac OS.

P

paraphrase   To use different words to express the same idea.

patience   The ability to bear trials calmly without complaint.

payroll tax   The employer's share of FICA taxes.

personnel manual   Document that outlines the employer's expectations and provides a basis of consistency in how practice issues are handled.

pharmaceuticals   Medicines including vaccines, antibiotics, anesthesia, sedatives, ointments, vitamins, and minerals.

physical inventory   Counting every single item in the practice and making a notation of the quantities on a list.

pitch   The highness or lowness of sound in a person's voice.

post   To transfer transactions from the daily journal to the accounts receivable ledger; when using a computer accounting program, it refers to entering the most recent transaction into the account.

practice manager   A person in a veterinary practice responsible for overseeing and coordinating the behind-the-scenes tasks and duties that allow the efficient delivery of medical care to patients and the accommodation of clients.

prejudice   Preconceived biases and opinions; everyone has prejudices.

printer   A piece of equipment that prints out text or images from the computer.

prioritize   To list tasks by importance, with the most important at the top and the least important at the bottom.

problem-oriented medical record   Organizing medical records in a detailed fashion for each patient.

procedures manual   Information about the regular events of the practice and policies toward patients.

professional courtesy   A tradition in the veterinary medical profession that discourages veterinarians from charging one another for veterinary medical treatment.

projection   When one's own ideas, feelings, or attitudes are attributed to someone else.

psychosomatic illness   Real physical symptoms resulting from an emotional or mental condition.

purge   To eliminate or delete medical records of animals that are no longer patients of the practice.

R

rationalization   Attributing one's actions to logical reasons.

receptionist   The person in a practice or hospital responsible for answering the phone, setting appointments, greeting clients as they enter, and any other front desk duties.

reference points   Factors that determine how messages are expressed and understood; may include education, experience, social and cultural barriers, and religious beliefs.

regression   Returning to an earlier mental or behavioral level during times of high stress.

relate   The facilitation of communication by having something in common with the other person.

repression   When socially unacceptable or painful desires or impulses are pushed out of the conscious mind into the unconscious, without the person being aware of it.

right-to-know station   Houses the practice's safety information.

rotate   To move products already on the shelves to the front and place newer items behind.

S

safety manual   Policies and instructions about minimizing the risks of accidents and injuries.

salary   The money earned by an employee on a weekly, monthly, or annual basis.

scanner   An imaging device that allows you to take a digital image of a document or photograph, and then transfers that image to the computer for editing, storage, or transmission.

screening   Handling incoming calls.

server   A computer or device that manages network resources.

sharps container   A heavy plastic container with a lid that can be permanently sealed, used for the safe disposal of hypodermic needles.

slander   A statement that harms a person's image or reputation, either intentionally or unintentionally.

SOAP   Progress notes in a problem-oriented medical record divided into Subjective, Objective, Assessment, and Procedure sections.

social security number   The number assigned to American citizens by the Social Security Administration. All employees must have a social security number.

software   Computer programs.

source-oriented medical record   Organizing medical records as acquired, in chronological order and for multiple patients.

squeeze chute   A small stall used to restrain large animals.

station   Individual unit of a call director telephone system, in which a call can be taken while other calls are coming in simultaneously.

statute of limitations   The amount of time in which a practice can take legal action against a client for nonpayment.

stereotype   Preconceived ideas about a group of people made without taking individual difference into account.

stress   The physical and/or psychological changes that occur in your body as a result of a change in your environment.

subjective data   Nonmeasurable data that describes an animal's attitude.

sublimation   Diverting an instinctual desire or impulse into a socially acceptable activity.

substance abuse   The overuse, inappropriate use, or illegal use of drugs, alcohol, or other potentially harmful substance.

T

tact Doing and saying the right things at the right time.

telephone reminder A telephone call to a client to remind him or her of a balance due on an account.

temporary withdrawal Finding ways to avoid dealing with a painful or difficult situation.

tickler file A file somewhat like an appointment book, with slots for mail and telephone messages, organized by days of the month.

time audit A method of time management that involves noting what you do throughout the day and how much time you spend doing it. This helps you to recognize time use patterns, which helps you develop a strategy for being more efficient with your time.

time tracking Keeping track of time as you work, in order to prevent yourself from spending too much time doing something to the exclusion of something else.

tone Vocal quality that expresses mood or feeling.

toner A special kind of dry, powdered ink used in laser printers and photocopiers. Toner generally comes in long cartridges specifically designed for individual printers or copiers.

V

vaccination certificate A formal document that allows a client to prove that an animal has been inoculated.

verbatim Word-for-word.

veterinarian An individual licensed to practice veterinary medicine or surgery; the member of the veterinary health team ultimately responsible for the administration of treatment, prescribing medication, performing surgery, and euthanasia.

veterinary assistant The person responsible for assisting the veterinary technician and veterinarian as needed; duties range from providing a limited degree of medical care (assisting in treatment, providing daily care to hospitalized patients, performing certain procedures, etc.) to administrative duties such as making appointments, to a hybrid of the two, such as filling prescriptions and educating clients.

veterinarian-client-patient relationship (VCPR) The relationship between the veterinarian and the client and patient.

veterinarian technician A trained professional who directly assists the veterinarian in the administration of health care to patients; the equivalent of a nurse in human health care.

volume Degree of loudness.

W

Wage and Tax Statement Form W-2; a statement of the total amount earned by an employee and amounts withheld for taxes.

want list Place to make notes of medications or supplies that have dropped to low quantities.

wave scheduling Scheduling the number of patients that can be seen in an hour's worth of segments at the same time, and then seeing patients on a first-come, first-served basis.

Web site A virtual location on the World Wide Web.

word processing program A special type of software that emulates a typewriter, although it generally includes additional features such as spell checking, automatic formatting, and the ability to cut and paste text and images.

workstation An individual computer on a network.

written reminder A written reminder on a statement or a letter to a client to remind him or her of a balance due on an account.

X

x-ray logbook Logbook used to record the use of x-ray equipment; contains the client's name, the patient's name, the date, x-ray case number, body part to be radiographed, the views taken, the measured thickness of the body parts radiographed, the x-ray machine settings, and any comments.

Z

zip drive A high-capacity storage device that can hold large quantities of data. The most current zip disks can hold up to 750mb (megabytes) of information.

Bibliography

Allen, Moira Anderson, M.Ed. Ten Tips on Coping with Pet Loss. 2002. Available at <http://www.pet-loss.net>. Accessed October 11, 2002.

Downing, Robin, D.V.M. *General Accounting Practices Study Unit*. New York: Delmar, 2000.

Downing, Robin, D.V.M. *Interpersonal Communication Study Unit*. New York: Delmar, 2000.

Downing, Robin, D.V.M. *Veterinary Practice and Administration Study Unit*. New York: Delmar, 2000.

Occupational Safety and Health Administration. Employer Responsibilities. 2002. Available at <http://www.osha.gov/> Search Terms: Employer Responsibility. Accessed October 11, 2002.

Sauter, Steven, Lawrence Murphy, et al. *Stress . . . at Work*. DHHS (NIOSH) Publication No. 99-101. 1999. Available at <http://www.cdc.gov/> Search Terms: Stress at Work. Accessed October 11, 2002.

Index